Uncovering
STUDENT THINKING
in
MATHEMATICS
Grades K-5

*To the many teachers and students who helped inform
the processes and resources in this book.
We will be forever grateful.*

Uncovering
STUDENT THINKING
in
MATHEMATICS
Grades K–5

25
Formative Assessment Probes *for the* Elementary Classroom

CHERYL ROSE TOBEY • LESLIE MINTON

Foreword by Page Keeley

CORWIN
A SAGE Company

For information:

Corwin
A SAGE Company
2455 Teller Road
Thousand Oaks, California 91320
(800) 233-9936
Fax: (800) 417-2466
www.corwinpress.com

SAGE Ltd.
1 Oliver's Yard
55 City Road
London EC1Y 1SP
United Kingdom

SAGE Pvt. Ltd.
B 1/I 1 Mohan Cooperative
 Industrial Area
Mathura Road, New Delhi 110 044
India

SAGE Asia-Pacific Pte. Ltd.
33 Pekin Street #02-01
Far East Square
Singapore 048763

Printed in the United States of America

Library of Congress Cataloging-in-Publication Data

Tobey, Cheryl Rose.
Uncovering student thinking in mathematics, grades K-5 : 25 formative assessment probes for the elementary classroom / Cheryl Rose Tobey, Leslie Minton.
     p. cm.
Includes bibliographical references and index.
ISBN 978-1-4129-8055-5 (pbk. : alk. paper)

  1. Mathematics—Study and teaching (Elementary) 2. Mathematics teachers—Training of—Handbooks, manuals, etc. I. Minton, Leslie. II. Title.

QA135.6.T6 2011
372.7—dc22   2010029355

This book is printed on acid-free paper.

14 15 16 17 18 10 9 8 7 6 5 4 3

Acquisitions Editor:	Jessica Allan
Associate Editor:	Joanna Coelho
Editorial Assistant:	Allison Scott
Production Editor:	Veronica Stapleton
Copy Editor:	Adam Dunham
Typesetter:	C&M Digitals (P) Ltd.
Proofreader:	Wendy Jo Dymond
Indexer:	Molly Hall
Cover Designer:	Scott Van Atta

Contents

Foreword   ix
Page Keeley

Preface   xi

Acknowledgments   xv

About the Authors   xvii

Chapter 1 Mathematics Assessment Probes   1

What Types of Understandings and Misunderstandings
   Does a Mathematics Assessment Probe Uncover?   2
How Were the Mathematics Assessment Probes Developed?   7
What Is the Structure of a Mathematics Assessment Probe?   10
What Additional Information Is Provided With
   Each Mathematics Assessment Probe?   17
What Mathematics Assessment Probes Are Included in the Book?   24

Chapter 2 Instructional Implications   29

Differentiating Instruction   30
Assessing Point of Entry   30
Analyzing Trends in Thinking   31
Giving Student Interviews   32
Promoting Math Conversations   33
Developing Vocabulary   34
Allowing for Individual Think Time   35
Improving Communication and Process Skills   36
Assessing Effectiveness of Instructional Activities and Materials   37
Building Capacity Across Grade Levels and Spans   38
Summary   40

**Chapter 3 Structure of Number Probes: Place Value,
Number Charts, and Number Lines**   41

How Many Stars?   42
Teacher Notes: How Many Stars?   43
Variation: How Many Counters?   48
What's the Number?   49

Teacher Notes: What's the Number? 50

Variation: What's the Number? 53

What Is the Value of the Place? 54

Teacher Notes: What Is the Value of the Place? 55

What Number Is That? 58

Teacher Notes: What Number Is That? 59

Variation: What Number Is That? 63

Hundred Chart Chunks 64

Teacher Notes: Hundred Chart Chunks 65

Variation: Hundred Chart Chunks 70

What Is the Value of the Digit? 71

Teacher Notes: What Is the Value of the Digit? 72

Variation: What Is the Value of the Digit? Card Sort 76

Chapter 4 Structure of Number Probes: Parts and Wholes and Equality 77

Equal to 4? 78

Teacher Notes: Equal to 4? 79

Variation A: Equal to 14? 83

Variation B: Equal to 14? 84

Crayon Count 85

Teacher Notes: Crayon Count 86

Variation: Crayon Count 90

Is $\frac{1}{4}$ of the Whole Shaded? 91

Teacher Notes: Is $\frac{1}{4}$ of the Whole Shaded? 92

Variation: Is $\frac{1}{4}$ of the Whole Shaded? Card Sort 97

Variation: How Much Is Shaded? 98

Granola Bar 99

Teacher Notes: Granola Bar 100

Is It Equivalent? 104

Teacher Notes: Is It Equivalent? 105

Variation: Is It Equivalent? Card Sort 109

Is It Simplified? 110

Teacher Notes: Is It Simplified? 111

Chapter 5 Structure of Number Probes: Computation and Estimation 115

How Many Dots? 116

Teacher Notes: How Many Dots? 117

Variation: How Many Counters? 121

Play Ball 122

Teacher Notes: Play Ball 123

Variation: Play Ball 127

What's Your Addition Strategy? 128

Teacher Notes: What's Your Addition Strategy? 129

Variation A: What's Your Addition Strategy? 134

Variation B: What's Your Addition Strategy? 135

What's Your Subtraction Strategy? 136

Teacher Notes: What's Your Subtraction Strategy?   137
Variation A: What's Your Subtraction Strategy?   142
Variation B: What's Your Subtraction Strategy?   143
What's Your Multiplication Strategy?   144
Teacher Notes: What's Your Multiplication Strategy?   145
Variation: What's Your Multiplication Strategy?   149
What's Your Division Strategy?   150
Teacher Notes: What's Your Division Strategy?   151
Is It an Estimate?   155
Teacher Notes: Is It an Estimate?   156
What Is Your Estimate?   160
Teacher Notes: What Is Your Estimate?   161

Chapter 6 Measurement, Geometry, and
**   Data Probes: Quadrilaterals   165**

Quadrilaterals   166
Teacher Notes: Quadrilaterals   167
Variation: Name that Shape   171
Variation: Is It a Polygon?   172
Variation: Is It a Circle?   173
What Is the Area?   174
Teacher Notes: What Is the Area?   175
Variation: What Is the Area?   179
What's the Measure?   180
Teacher Notes: What's the Measure?   181
Variation: Length of Line   185
Variation: What's the Measure?   186
Graph Choices   187
Teacher Notes: Graph Choices   188
Variation: Name the Graph   192
Variation: What Does the Graph Say?   193
The Median   194
Teacher Notes: The Median   195
Variation: Effect on the Median Card Sort   198

Resource: Notes Template: QUEST Cycle   199

References   201

Index   205

Foreword

Probe- n. 1. A slender surgical instrument for exploring a wound, 2. investigation, 3. Spacecraft, etc. used to get information about an environment- v. 1) explore with a probe, 2) investigate- **prob'er** *n.*

—*Webster's New Pocket Dictionary* (2000)

Teachers face the challenging and complex task of linking important ideas in mathematics, research about how students learn, and the myriad thinking styles students bring to the classroom. How do we bring together these three critical components—content, cognitive research, and our own students' thinking—in a way that informs teaching and supports learning for all students? The solution lies in the unique two-tiered diagnostic and formative assessment probes, along with supporting teacher background material, provided in the chapters that follow. This book instructs teachers to systematically use formative assessment probes as they build and refine their use of assessment *for* learning to answer the pivotal question, *what and how are my students thinking in relation to the mathematics I am teaching?*

The first question you might have asked when you read the title of this book is, what is a "probe"? According to the *Webster's* definition above, it is an instrument that can be used to explore a health problem in the body. In the context of this book, it's a diagnostic instrument that allows us to explore student thinking in order to understand a learning problem. *Webster's* also defines a probe as an investigation. In mathematics, we use probes to investigate the ideas and modes of reasoning students use to make sense of mathematical problems. Furthermore, *Webster's* describes a probe as a tool that gives us information about the environment. Probes provide teachers with information about the effectiveness of the teaching and learning environment so that instructional strategies can be matched to students' learning needs. And finally, *Webster's* indicates that "probe" can also be used as a verb. When we probe in mathematics, we are exploring students' thinking and becoming investigators of learning in our classrooms. In essence, the probes in this book are specially designed questions that enable teachers to identify the ideas students bring to their learning and use that knowledge inform their teaching. A thoughtful analysis of students' ideas can help teachers make informed decisions about the instructional path needed to move them from where they are in their conceptual and procedural understandings to the destination of mathematical literacy.

Over the past several years, I have had the privilege of working with Cheryl Rose Tobey on two National Science Foundation–funded projects that developed parallel formative assessment resources for both science and mathematics. Although our work focused on different disciplines, our philosophies and approaches to formative assessment have been very similar. We both believe that if teachers are to use formative assessment results effectively, the strategies they employ must be linked to explicit learning targets. In addition, teachers need to be familiar with learning research that reveals the difficulties students may encounter in accessing content and the conceptual barriers that may impede their learning. Once teachers are grounded in the content, standards, and research on learning, they can begin the systematic cycle of collecting data, analyzing student thinking, and choosing strategies that address students' needs in the classroom. In the process, teachers come to realize that these important formative assessment tools also contribute to their ongoing professional development, deepening their pedagogical content knowledge. This book belongs in the professional library of every elementary school mathematics teacher.

As a science educator and author of several books on science formative assessment, I am thrilled to recommend this book to my science colleagues as well as to teachers of mathematics. With the emphasis on Science, Technology, Engineering, and Mathematics (STEM) education increasingly finding its way into the elementary classroom, teachers are looking for ways to meaningfully integrate mathematics and science. Many of the probes in this book address important inquiry skills students use in science such as measurement, estimation, representing and analyzing data, performing computations, choosing appropriate quantitative data, and more. As I work with K–12 science teachers throughout the United States in using assessment probes, I am frequently asked if there are similar probes in mathematics that can clarify the difficulties students encounter in applying mathematical concepts and procedures to science. And now my answer is a resounding *yes*—in this book! We now have a full suite of K–12 mathematics formative assessment probes that can be combined with their science counterparts to support and transform assessment, instruction, student learning, and teacher professional development.

In the busy day-to-day life of a classroom, we must all take the time to listen to our students and provide opportunities for them to share their thinking. Eleanor Duckworth of Harvard University said in her classic book, *The Having of Wonderful Ideas and Other Essays on Teaching and Learning*, "The having of wonderful ideas is what I consider the essence of intellectual development." (1996, p. 1) Thank you to Cheryl Rose Tobey and Leslie Minton for giving us another book designed to reveal those wonderful ideas students have that are sure to promote learning and foster a love of mathematics!

Page Keeley, Maine Mathematics and Science Alliance
Past President of the National Science Teachers Association (NSTA)

Preface

OVERVIEW

With mandates from No Child Left Behind and other state-driven assessment initiatives, substantial educator time and energy are being spent on developing, implementing, scoring, and analyzing summative assessments of students' mathematical knowledge and analyzing results from comprehensive diagnostic assessment systems covering a range of topics in a given setting. Although the importance of summative and large-scale diagnostic assessment is recognized, findings point to formative assessment as an important strategy in improving student achievement in mathematics.

Formative assessment informs instruction through varying methods and strategies, the purposes of which are to determine students' prior knowledge of a learning target and to use the information to drive instruction, moving each student toward understanding of the targeted concepts and procedures. Questioning, observation, and student self-assessment are examples of instructional strategies educators can incorporate to gain insight into student understanding. These instructional strategies become formative assessments if the results are used to plan and implement learning activities designed specifically to address the specific needs of the students.

This book focuses on using short sets of diagnostic questions, called *Mathematics Assessment Probes.* The probes can elicit prior understandings and commonly held misconceptions. This elicitation allows the educator to make sound instructional choices, targeted to a specific mathematics concept, based on the specific needs of a particular group of students.

> Diagnostic assessment is as important to teaching as a physical exam is to prescribing an appropriate medical regimen. At the outset of any unit of study, certain students are likely to have already mastered some of the skills that the teacher is about to introduce, and others may already understand key concepts. Some students are likely to be deficient in prerequisite skills or harbor misconceptions. Armed with this diagnostic information, a teacher gains greater insight into what to teach. (McTighe & O'Connor, 2005, page 14)

The Mathematics Assessment Probes provided here are tools for elementary school teachers to gather important insights in a practical way, providing immediate information for planning purposes.

AUDIENCE

The first collection of Mathematics Assessment Probes and the accompanying Teacher Notes was designed for the busy K–12 classroom teacher who understands there is a growing body of research on students' learning difficulties and that thoughtful use of this research in developing and selecting diagnostic assessments promises to enhance the efficiency and effectiveness of mathematics instruction. Since the publication of the collection, *Uncovering Student Thinking in Mathematics: 25 Formative Assessment Probes* (Rose, Minton, & Arline, 2007), we have received continuous requests for additional probes. Both teachers and education leaders have communicated the need for a collection of research-based probes that focuses on a narrower grade span. Due to these requests, we set to work writing, piloting, and field testing a more extensive set of probes for primary, intermediate, middle and high school teachers. As a result, *Uncovering Student Thinking in Mathematics: 30 Formative Assessment Probes for the Secondary Classroom* (Rose & Arline, 2009) was written and published, and our full attention was then turned to finalizing this collection of probes designed for the elementary setting.

BACKGROUND

The probes are designed to uncover student understandings and misunderstandings based on research findings, and they have been piloted and field tested with teachers and students.

Because the probes are based on cognitive research, examples of such probes exist in multiple resources but not as a collection and not for the specific purpose of action research in the classroom. In addition, the questions are spread throughout various research materials and are not ready for classroom use. The probes in this book were developed using the process described in *Mathematics Curriculum Topic Study: Bridging the Gap Between Standards and Practice* (Keeley & Rose, 2007) and were originally piloted in Maine. The use of the probes was expanded to include participants in the other mathematics projects, including the National Science Foundation–funded Northern New England Co-Mentoring Network, Maine Governor's Academy for Mathematics and Science Education Leadership, the state-funded Early Mathematics Thinking Project, the State Mathematics and Science Partnership Project: Building Administrators' and Leaders' Abilities and the Numeracy Capacity of Educators, and various other mathematics professional development programs offered through the Maine Mathematics and Science Alliance. In addition, the probes were field-tested nationally through a network of leaders involved with the Curriculum Topic Study Project. While many of these projects were no longer current during the development of this resource, many of the past participants have continued to play a critical role in helping us pilot and field-test the newly developed probes.

ORGANIZATION

This book is organized to provide readers with the purpose, structure, and development of the Mathematics Assessment Probes as well as to support the use of applicable research and instructional strategies in mathematics classrooms.

Chapter 1 provides in-depth information about the process and design of the Mathematics Assessment Probes along with the development of an action-research structure we refer to as a QUEST cycle. Chapter 2 highlights instructional implications and images from practice to illuminate how easily and in how many varied ways the probes can be used in mathematics classrooms. Chapters 3 through 6 are collections of probes categorized by concept strands with accompanying Teacher Notes that provide the specific research and instructional strategies designed to directly address students' challenges with mathematics.

Acknowledgments

We would like to thank the many mathematics educators who gave valuable feedback about various features of the probes, including structures, concepts to target, and purposes of use. This group includes teachers and administrators from schools in Nashua, New Hampshire; Springfield, Illinois; Brookline, New Hampshire; Sullivan, Maine; Colorado Springs area, Colorado; Waterville, Maine; Auburn, Maine; Kennebunkport, Maine, Wells, Maine; and Maine districts SAD 57, SAD 11, SAD 22, and SAD 72. In addition to the schools and district groups, we received valuable information from the many Curriculum Topic Study session participants who met to learn the process of designing research-based diagnostic probes.

We thank the many educators who field-tested the probes and would especially like to acknowledge the contributions of the following educators for supplying sample student work, ideas for probes, feedback for teacher notes and/or allowing us to interview students: Nancy Barrett, Celeste Beaulieu, Michele Chadburn, Debbie Cook, Christine Downing, Tracey Hartnett, Andrea Kutinsky, Kristin Larrabee, Jennifer Larrenga, Denise Masalsky, Christina McGowan, Priscilla McFarland, Rebecca Morse, Karen Pillion, Marcia Tobey, Sue Williamson, Mary Wilson, and Sarah Young.

Again, we would like to thank Page Keeley, our science colleague, who designed the process for developing diagnostic assessment probes and who tirelessly promotes the use of formative assessments, helping to disseminate our work in her travels. Thanks to Pam Buffington for her continual encouragement; and our sincere gratitude to Carolyn Arline, who helped review content and is always available to talk about mathematics.

In addition, we would like to acknowledge the continued support of our families during the book-writing process. A special thanks to Samantha and Jack, who were always willing to be the first to try out a newly developed probe.

PUBLISHER'S ACKNOWLEDGMENTS

Corwin gratefully acknowledges the contributions of the following reviewers:

Carol Amos
Teacher Leader/Mathematics Coordinator
Twinfield Union School
Plainfield, Vermont

Virginia M. Hughes
Mathematics Instructor
Washington State University
Pullman, Washington

Renee Peoples
Teacher, K–5 Math Facilitator
Swain County Schools
Bryson City, North Carolina

Pearl G. Solomon
Professor Emeritus of Education
St Thomas Aquinas College
Sparkill, New York

Randy Wormald
Math Teacher
Belmont High School
Belmont, New Hampshire

About the Authors

Cheryl Rose Tobey is a project director for the Northeast and the Islands Regional Educational Laboratory housed at the Education Development Center (EDC). Her work is primarily in the areas of leadership, mathematics professional development, and school reform. Before joining EDC, Tobey was the senior program director for mathematics at the Maine Mathematics and Science Alliance (MMSA) where she served as the coprincipal investigator of the mathematics section of the National Science Foundation (NSF)-funded project, Curriculum Topic Study, and principal investigator and project director of two Title IIa State Mathematics and Science Partnership projects. Prior to working on these projects, Tobey was the coprincipal investigator and project director for MMSA's NSF-funded Local Systemic Change Initiative, Broadening Educational Access to Mathematics in Maine, and she was a fellow in Cohort 4 of the National Academy for Science and Mathematics Education Leadership. Before joining MMSA in 2001, Tobey was a high school and middle school mathematics educator for 10 years. She received her BS in secondary mathematics education from the University of Maine at Farmington and her MEd from City University in Seattle.

Leslie Minton, a mathematics consultant for Math Matters 2, Portland, Maine, is currently providing individualized mathematics professional development to area schools and districts Grades PK–8. As well as providing mathematics professional development, Leslie teaches math methods courses at the University of Southern Maine. Before starting Math Matters 2, Leslie was a project director for the Maine Mathematics and Science Alliance, Augusta, Maine. She provided technical assistance to schools as well as created a professional development course and diagnostic materials designed to support numeracy understanding. She is a fellow of the second cohort group of the Governor's Academy for Science and Mathematics Educators and has taught regular and special education for Grades 4 through 12. Leslie received her BS in elementary and special education from the University of Maine at Farmington and her MEd in curriculum, instruction, and assessment from Walden University. She recently completed an MEd program in educational design and media technology at Full Sail University.

Mathematics Assessment Probes

To differentiate instruction effectively, teachers need diagnostic assessment strategies to gauge their students' prior knowledge and uncover their understandings and misunderstandings. By accurately identifying and addressing misunderstandings, teachers prevent their students from becoming frustrated and disenchanted with mathematics, which can reinforce the student preconception that "some people don't have the ability to do math." Diagnostic strategies also allow for instruction that builds on individual students' existing understandings while addressing their identified difficulties. The Mathematics Assessment Probes in this book allow teachers to target specific areas of difficulty as identified in research on student learning. Targeting specific areas of difficulty—for example, the transition from reasoning about whole numbers to understanding numbers that are expressed in relationship to other numbers (decimals and fractions)—focuses diagnostic assessment effectively (National Research Council, 2005, p. 310).

Mathematics Assessment Probes represent one approach to diagnostic assessment. The probes specifically elicit prior understandings and commonly held misconceptions that may or may not have been uncovered during an instructional unit. This elicitation allows teachers to make instructional choices based on the specific needs of students. Examples of commonly held misconceptions elicited by a Mathematics Assessment Probe include ideas such as "an equals sign means *the answer follows*" and "to add fractions, add the numerators and then add the denominators." It is important to make the distinction between what we might call a silly mistake and a more fundamental one, which may be the product of a deep-rooted misunderstanding. It is not uncommon for different students to display the same misunderstanding every year. Being aware of and eliciting common misunderstandings and drawing students' attention to them can be a valuable teaching technique (Griffin & Madgwick, 2005) that should be used no matter what particular curriculum program a teacher uses, be it commercial, district developed, or teacher developed.

The process of diagnosing student understandings and misunderstandings and making instructional decisions based on that information is the key to increasing students' mathematical knowledge.

To use the Mathematics Assessment Probes for this purpose, teachers need to

- determine a question;
- use a probe to examine student understandings and misunderstandings;
- use links to cognitive research, standards, and math education resources to drive next steps in instruction;
- implement the instructional unit or activity; and
- determine the impact on learning by asking an additional questions.

The probes and the above process are described in detail in this chapter. The Teacher Notes that accompany each of the Mathematics Assessment Probes in Chapters 3 through 6 include information on research findings and instructional implications relevant to the instructional cycle described above. Detailed information about the information provided in the accompanying Teacher Notes is also described in detail within this chapter.

WHAT TYPES OF UNDERSTANDINGS AND MISUNDERSTANDINGS DOES A MATHEMATICS ASSESSMENT PROBE UNCOVER?

Developing understanding in mathematics is an important but difficult goal. Being aware of student difficulties and the sources of those difficulties, and designing instruction to diminish them, are important steps in achieving this goal (Yetkin, 2003). The Mathematics Assessment Probes are designed to uncover student understandings and misunderstandings; the results are used to inform instruction rather than make evaluative decisions. As shown in Figure 1.1, the understandings include both conceptual and procedural knowledge, and misunderstandings can be classified as common errors or overgeneralizations. Each of these is described in the following in more detail.

Understandings: Conceptual and Procedural Knowledge

Research has solidly established the importance of conceptual understanding in becoming proficient in a subject. When students understand mathematics, they are able to use their knowledge flexibly. They combine factual knowledge, procedural facility, and conceptual understanding in powerful ways. (National Council of Teachers of Mathematics [NCTM], 2000)

Figure 1.1   Diagnostic Assessment Probes

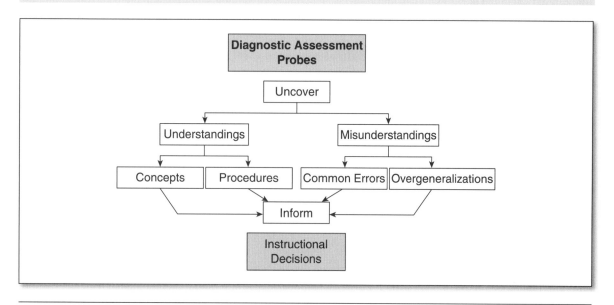

Source: Rose, C., Minton, L. & Arline, C. (2007).

Conceptual Understanding

Students demonstrate conceptual understanding in mathematics when they

- recognize, label, and generate examples and nonexamples of concepts;
- use and interrelate models, diagrams, manipulatives, and so on;
- know and apply facts and definitions;
- compare, contrast, and integrate concepts and principles;
- recognize, interpret, and apply signs, symbols, and terms; and
- interpret assumptions and relationships in mathematical settings.

Procedural Knowledge

Students demonstrate procedural knowledge in mathematics when they

- select and apply appropriate procedures;
- verify or justify a procedure using concrete models or symbolic methods;
- extend or modify procedures to deal with factors in problem settings;
- use numerical algorithms;
- read and produce graphs and tables;
- execute geometric constructions; and
- perform noncomputational skills, such as rounding and ordering.

Source: From U.S. Department of Education, 2003, Chapter 4.

The relationship between understanding concepts and being proficient with procedures is complex. The following description gives an example of how the Mathematics Assessment Probes elicit conceptual or procedural understanding. The What Is the Value of the Digit? probe (see Figure 1.2) is designed to elicit whether students understand place value beyond being able to procedurally connect numbers to their appropriate places.

Students who choose B, There is a 2 in the ones place, and E, There is a 1 in the tenths place, understand the value of the place of digits within a number. By also choosing C, There are 21.3 tenths, and H, There are 213 hundredths, these students also demonstrate a conceptual understanding of the relationship between the value of the places and the number represented by the specific combination of digits.

Figure 1.2   What Is the Value of the Digit? (see page 71, Probe 6, for more information on this probe)

Teacher prompt: "Circle all of the statements that are true for the number 2.13."

Statement	Explanation (why circled or not circled)
A) There is a 3 in the ones place.	
B) There is a 2 in the ones place.	
C) There are 21.3 tenths.	
D) There are 13 tenths.	
E) There is a 1 in the tenths place.	
F) There is a 3 in the tenths place.	
G) There are 21 hundredths.	
H) There are 213 hundredths.	

Misunderstandings: Common Errors and Overgeneralizations

In *Hispanic and Anglo Students' Misconceptions in Mathematics*, Jose Mestre (1989) describes misconceptions as follows:

Students do not come to the classroom as "blank slates." Instead, they come with theories constructed from their everyday experiences. They have actively constructed these theories, an activity crucial to all successful learning. Some of the theories that students use to make sense of the world are, however, incomplete half-truths. They are misconceptions.

Misconceptions are a problem for two reasons. First, they interfere with learning when students use them to interpret new experiences. Second, students are emotionally and intellectually attached to their misconceptions because they have actively constructed them. Hence, students give up their misconceptions, which can have such a harmful effect on learning, only with great reluctance. (para. 2–3)

For the purposes of this book, these misunderstandings or misconceptions will be categorized into common errors and overgeneralizations, which are described in more detail below.

Common Error Patterns

Common error patterns refer to systematic uses of inaccurate and inefficient procedures or strategies. Typically, this type of error pattern indicates nonunderstanding of an important math concept (University of Kansas, 2005). Examples of common error patterns include consistent misuse of a tool or steps of an algorithm, such as an inaccurate procedure for computing or the misreading of a measurement device. The following description gives an example of how the Mathematics Assessment Probes elicit common error patterns.

One of the ideas the What's the Measure? probe (see Figure 1.3) is designed to elicit is the understanding of zero point. "A significant minority of older children (e.g., fifth grade) respond to nonzero origins by simply reading off

Figure 1.3   What's the Measure? Probe (See page 180, Probe 23, for more information on this probe.)

Use the measuring tool provided to measure the length of the line.

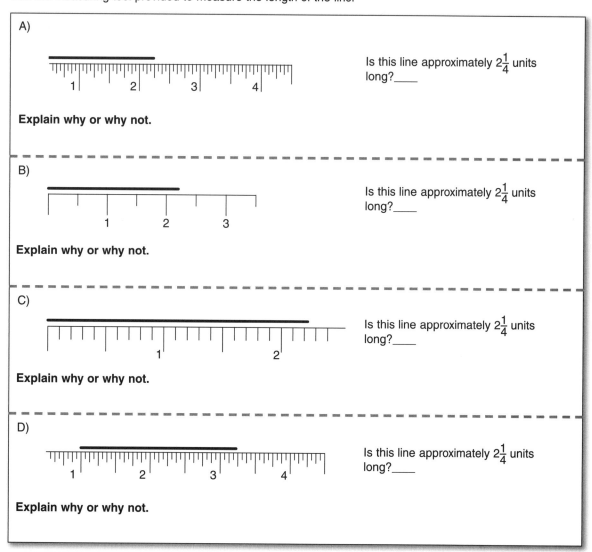

whatever number on a ruler aligns with the end of the object (Lehrer et al., 1998a)" (NCTM, 2003, p. 183). We have found many middle school students also make this same mistake.

The correct answers are A, no; B, yes; C, yes; and D, yes. Students who *include* A typically do not consider the nonzero starting point and give the length as the number on the ruler aligned to the endpoint of the segment. Students who exclude D are not considering the nonzero starting point.

Overgeneralizations

Often, students learn an algorithm, rule, or shortcut and then extend this information to another context in an inappropriate way. These misunderstandings are often overgeneralizations from cases that students have seen in prior instruction (Griffin & Madgwick, 2005). To teach in a way that avoids creating any misconceptions is not possible, and we have to accept that students will make some incorrect generalizations that will remain hidden unless the teacher makes specific efforts to uncover them (Askew & Wiliam, 1995). The following example illustrates how the Mathematics Assessment Probes can elicit overgeneralizations. The What is the Number? probe (see Figure 1.4) is designed to elicit the overgeneralizations students make from the way a number

Figure 1.4   What's the Number? (See page 49, Probe 2, for more information on this probe.)

Teacher prompt: "Circle the calculator that shows the number *one hundred thirty-two.*"

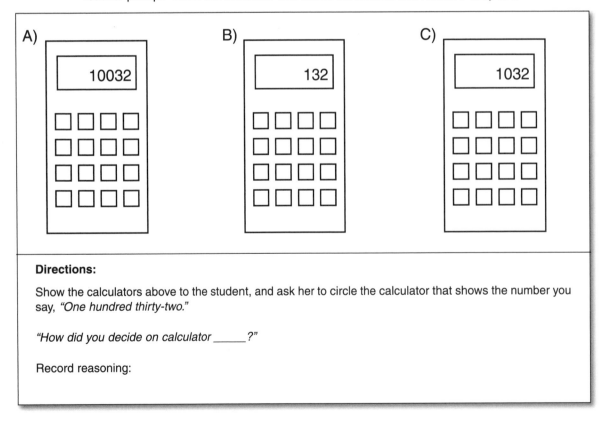

Directions:

Show the calculators above to the student, and ask her to circle the calculator that shows the number you say, *"One hundred thirty-two."*

"How did you decide on calculator _____?"

Record reasoning:

is read orally to the numeric form. Students who incorrectly choose A often disregard place value concepts because of this overgeneralization.

A single probe can elicit both common errors and overgeneralizations depending on how individual students respond to the particular question. In addition to uncovering common misunderstandings, the Mathematics Assessment Probes also elicit *uncommon* misconceptions that may not be uncovered and could continue to cause difficulty in understanding a targeted concept.

HOW WERE THE MATHEMATICS ASSESSMENT PROBES DEVELOPED?

Developing an assessment probe is different from creating appropriate questions for summative quizzes, tests, or state and national exams. The probes in this book were developed using the process described in *Mathematics Curriculum Topic Study: Bridging the Gap Between Standards and Practice* (Keeley & Rose, 2006).

The process is summarized as follows:

- **Identify the topic you plan to teach, and use national standards to examine concepts and specific ideas related to the topic.** The national standards used to develop the probes for this book were NCTM's (2000) *Principles and Standards for School Mathematics* and the American Association for the Advancement of Science's (AAAS, 1993) *Benchmarks for Science Literacy*.

- **Select the specific concepts or ideas you plan to address, and identify the relevant research findings.** The source for research findings include NCTM's (2003) *Research Companion to Principles and Standards for School Mathematics*, Chapter 15 of AAAS's (1993) *Benchmarks for Science Literacy*, and additional supplemental articles related to the topics of the probes.

- **Focus on a concept or a specific idea you plan to address with the probe, and identify the related research findings.** Choose the type of probe structure that lends itself to the situation (see more information on probe structure following the Gumballs in a Jar example on page 9). Develop the stem (the prompt), key (correct response), and distracters (incorrect responses derived from research findings) that match the developmental level of your students.

- **Share your assessment probes with colleagues for constructive feedback, pilot with students, and modify as needed.**

Figure 1.5, which is taken from Keeley and Rose (2006), provides the list of concepts and specific ideas related to the probability of simple events. The shaded information was used as the focus in developing the probe Gumballs in a Jar (see Figure 1.6).

Figure 1.5 Probability Example

Topic: Probability (Simple Events)

Concepts and Ideas	Research Findings
• Events can be described in terms of Being more or less likely, impossible or certain (Grades 3–5, AAAS, 1993, p. 228). • Probability is the measure of the likelihood of an event and can be represented by a number from 0 to 1 (Grades 3–5, NCTM, 2000, p. 176). • Understand that 0 represents the probability of an impossible event and 1 represents the probability of a certain event (Grades 3–5, NCTM, 2000, p. 181). • Probabilities are ratios and can be expressed as fractions, percentages, or odds (Grades 6–8, AAAS, 1993, p. 229). • Methods such as organized lists, tree diagrams, and area models are helpful in finding the number of possible outcomes (Grades 6–8, NCTM, 2000, pp. 254–255). • The theoretical probability of a simple event can be found using the ratio of a favorable outcome to total possible outcomes (Grades 6–8, AAAS, 1993, p. 229). • The probability of an outcome can be tested with simple experiments and simulation (NCTM, 2000, pp. 254–255). • The relative frequency (experimental probability) can be computed using data generated from an experiment or simulation (Grades 6–8, NCTM, 2000, pp. 254–255). • The experimental and theoretical probability of an event should be compared with discrepancies between predictions and outcomes from a large and representative sample taken seriously (Grades 6–8, NCTM, 2000, pp. 254–255).	**Understandings of Probability (NCTM, 2003, pp. 216–223)** • Lack of understanding of ratio leads to difficulties in understanding of chance. • Students tend to focus on absolute rather than relative size. • Although young children do not have a complete understanding of ratio, they have some intuitions of chance and randomness. • A continuum of probabilistic thinking includes subjective, transitional, informal, quantitative, and numerical levels. • Third grade (approx) is an appropriate place to begin systematic instruction. • "Equiprobability" is the notion that all outcomes are equally likely, disregarding relative and absolute size. • The outcome approach is defined as the misconception of predicting the outcome of an experiment rather than what is likely to occur. A typical response to questions is, "Anything can happen." • Intuitive reasoning may lead to incorrect responses. Categories include representativeness and availability. • Wording to task may influence reasoning. NAEP results show fourth and eighth graders have difficulty with tasks involving probability as a ratio of "m chances out of n" but not with "1 chance out of n" (NCTM, 2003, p. 222). • Increased understanding of sample space stems from multiple opportunities to determine and discuss possible outcomes and predict and test using simple experiments. **Uncertainty (AAAS, 1993, p. 353)** • Upper-elementary students can give correct examples for certain, possible and impossible events, but they have difficulties calculating the probability of independent and dependent events. • Upper-elementary students create "part to part" rather than "part to whole" relationships.

From Keeley, P., & Rose, C. M. (2006).

Figure 1.6 Gumballs in a Jar

Two jars both contain black and white gumballs.

Jar A: 3 black and 2 white

Jar B: 6 black and 4 white

Which response best describes the chance of getting a black gumball?

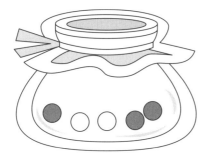 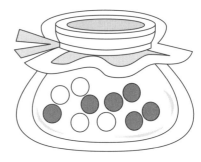

A. There is a better chance of getting a black gumball from Jar A.

B. There is a better chance of getting a black gumball from Jar B.

C. The chance of getting a black gumball is the same for both Jars A and B. Explain your reasons for the answer you selected.

_____

_____

_____

_____

_____

_____

_____

_____

_____

_____

_____

_____

_____

_____

_____

_____

_____

From Keeley, P., & Rose, C. M. (2006).

The probe is used to reveal common errors regarding probability, such as focusing on absolute size, or a lack of conceptual understanding of probability as a prediction of what is likely to happen. There is the same chance you will pick a black gumball out of each jar. Jar A has a probability of $\frac{3}{5}$, and Jar B has a probability of $\frac{6}{10} = \frac{3}{5}$. There are a variety of trends in correct thinking related to this probe, some of which are doubling, ratios, and percents. Some students might correctly choose answer C but use incorrect reasoning, such as "You can't know for sure since anything can happen," an explanation that indicates a lack of conceptual understanding of probability. Other students may demonstrate partial understanding with responses such as "each jar has more black than white." Some students reason that there are fewer white gumballs in Jar A compared to Jar B and therefore there is a better chance of picking a black gumball from Jar A. Others observe that Jar B has more black gumballs compared to Jar A and therefore reason that there is a better chance of picking a black gumball. In both cases, students are focusing on absolute size instead of relative size in comparing the likelihood of events. Students sometimes choose Distracter A due to an error in counting or calculation.

Additional probes can be written using the same list of concepts and specific ideas related to the probability of simple events. For example, by focusing on the statement from the research, "NAEP results show fourth and eighth graders have difficulty with tasks involving probability as a ratio of 'm chances out of n' but not with '1 chance out of n'" (NCTM, 2003, p. 222), a probe using an example of each can diagnose if students are demonstrating this difficulty.

WHAT IS THE STRUCTURE OF A MATHEMATICS ASSESSMENT PROBE?

A probe is a cognitively diagnostic paper-and-pencil assessment developed to elicit research-based misunderstandings related to a specific mathematics topic. The individual probes are designed to be (1) *easy to use* and copy ready for use with students; (2) *targeted* to one mathematics topic for short-cycle intervention purposes; and (3) *practical*, with administration time targeted to approximately 5 to 15 minutes.

Each one-page probe consists of selected response items (called Tier 1) and explanation prompts (called Tier 2), which together elicit common understandings and misunderstandings. Each of the tiers is described in more detail below.

Tier 1: Elicitation

As the elicitation tier is designed to uncover common understandings and misunderstandings, a structured format using a question or series of questions followed by correct answers and incorrect answers (often called distracters) is used to narrow ideas to those found in the related cognitive research. The formats typically fall into one of seven categories.

1. Selected response. One question, one correct answer, and several distracters (see Figures 1.7 and 1.8).

Figure 1.7 Variation: Crayon Count (See page 90, Probe 8A, for more information about this probe.)

There is a box of orange, blue, and purple crayons. Half ($\frac{1}{2}$) of the crayons are purple. Some crayons are orange, and a quarter ($\frac{1}{4}$) of the crayons are blue.

How many of the crayons are orange?

A) $\frac{3}{4}$

B) $\frac{2}{4}$

C) $\frac{1}{4}$

Figure 1.8 Variation: How Much Is Shaded? (See page 98, Probe 9B, for more information about this probe.)

What fraction of the shape is shaded?

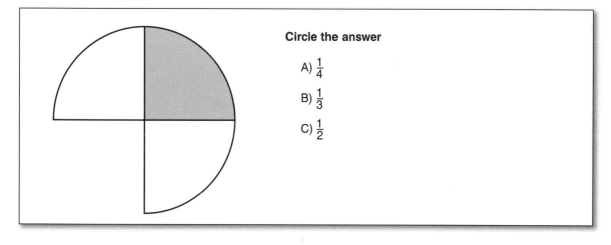

Circle the answer

A) $\frac{1}{4}$

B) $\frac{1}{3}$

C) $\frac{1}{2}$

2. Multiple selections response. Two or more sets of problems, each with one question, one correct answer, and one or more distracters per problem (see Figure 1.9).

Figure 1.9   What's the Area? (See page 174, Probe 22, for more information about this probe.)

Item	Select Answer
A) Explain your thinking:	**Area of Rectangle?** a) 12 sq units b) 6 sq units c) 9 sq units d) 5 sq units e) Not enough Information to find area
B) 2 2 8 2 2 11 Explain your thinking:	**Area of the Figure?** a) 88 sq units b) 27 sq units c) 38 sq units d) Not enough Information to find area
C) Area of Triangle = 7 sq units Explain your thinking:	**Area of Rectangle?** a) 49 sq units b) 14 sq units c) 28 sq units d) 21 sq units e) Not enough Information to find area

3. Open response. One or more sets of items, each with one question. The open-response format does not include Tier 1 selected response choices on the student probe (see Figure 1.10). Typical incorrect responses are provided in the teacher's notes rather than listed as distracters.

Figure 1.10   How Many Stars? (See page 42, Probe 1, for more information about this probe.)

(Student Interview Task)

Student writes answer on line.

How many stars are there?

_____

- -

Follow the directions below once student writes response.

Teacher points to the digit in the ones place of the student response and says,

"Can you use this *red* pencil to circle how many stars this number means from your answer?"

Next, point to the digit in the tens place of the student response and say,

Now use this *blue* pencil to circle how many stars this number means from your answer.

4. Opposing views/answers. Two or more statements are given, and students are asked to choose the statement they agree with (see Figure 1.11). This format is adapted from *Concept Cartoons in Science Education*, created by Stuart Naylor and Brenda Keogh (2000), for probing student ideas in science.

Figure 1.11 Quadrilaterals (See page 166, Probe 21, for more information on this probe.)

Circle the name or names of the people you agree with.

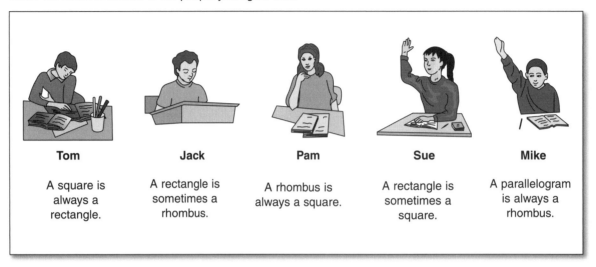

Tom	Jack	Pam	Sue	Mike
A square is always a rectangle.	A rectangle is sometimes a rhombus.	A rhombus is always a square.	A rectangle is sometimes a square.	A parallelogram is always a rhombus.

5. Examples and nonexamples list. One question or statement with several examples and nonexamples pertaining to a statement listed below. Students are asked to find only the examples based on a given statement (see Figure 1.12). This probe structure is often set up as a card sort where students are given one problem per card and asked to sort the problems into 2 piles.

Figure 1.12 Equal to 4? (See page 78, Probe 7, for more information on this probe.)

Circle only the math sentences where □ = 4.

A) 2 + 2 = □ − 3	D) □ = 1 + 3
B) 9 − □ = 5	E) 6 + 3 = □ + 5
C) 10 − 6 = □	F) 3 + 1 = □ + 2

6. Justified list. Two or more separate questions or statements are given, and students are asked to explain each choice.

Figure 1.13 Is It Equivalent? (See page 104, Probe 11, for more information on this probe.)

Equivalent to $\frac{2}{5}$?		Explain why or why not.
A) 2.5	Yes No	
B) 25%	Yes No	
C) 0.4	Yes No	
D) 0.25	Yes No	
E) 40%	Yes No	
F) 2.5%	Yes No	
G) 0.04	Yes No	

7. Strategy elicitation. A problem is stated with multiple solution strategies given. Students provide an explanation regarding making sense of each strategy.

Tier 2: Elaboration

The second tier of each of the probes is designed for individual elaboration of the reasoning used to respond to the question asked in the first tier. Mathematics teachers gain a wealth of information by delving into the thinking behind students' answers not just when answers are wrong but also when they are correct (Burns, 2005). Although the Tier 1 answers and distracters are designed around common understandings and misunderstandings, the elaboration tier allows educators to

Figure 1.14 What's Your Multiplication Strategy? (See page 144, Probe 17, for more information on this probe.)

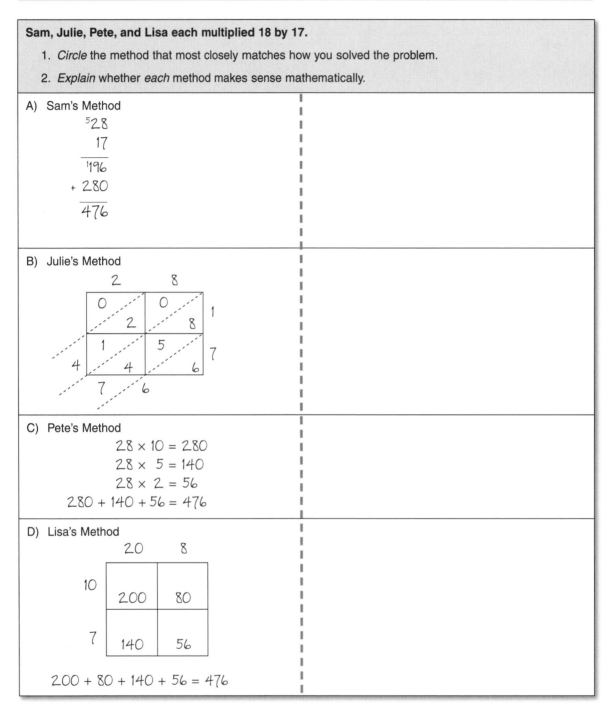

look more deeply at student thinking as sometimes a student chooses a specific response, correct or incorrect, for an atypical reason. Also, there are many different ways to approach a problem correctly; therefore, the elaboration tier allows educators to look for trends in thinking and in methods used.

Also important to consider is the idea that in order to address misconceptions, students must be confronted with their own incorrect ideas by participating in instruction that causes cognitive dissonance between existing ideas and new ideas. By having students complete both tiers of a probe and then planning instruction that addresses the identified areas of difficulty, teachers can then use students' original responses as part of a reflection on what was learned. Without this pre-assessment commitment of selecting an answer and explaining the choice, new understanding and corrected ideas are not always evident to the student.

WHAT ADDITIONAL INFORMATION IS PROVIDED WITH EACH MATHEMATICS ASSESSMENT PROBE?

In *Designing Professional Development for Teachers of Science and Mathematics,* Loucks-Horsley, Love, Stiles, Mundry, and Hewson (2003) describe action research as an effective professional development strategy. To use the probes in this manner, it is important to consider the complete implementation process.

We refer to an action research quest as working through the full cycle of

- questioning student understanding of a particular concept;
- uncovering understandings and misunderstandings using a probe;
- examining student work;
- seeking links to cognitive research to drive next steps in instruction; and
- teaching implications based on findings and determining impact on learning by asking an additional question.

The Teacher Notes, included with each probe, have been designed around the action research QUEST cycle, and each set of notes includes relevant information for each component of the cycle (see Figure 1.15). These components are described in detail below.

Figure 1.15   QUEST Cycle

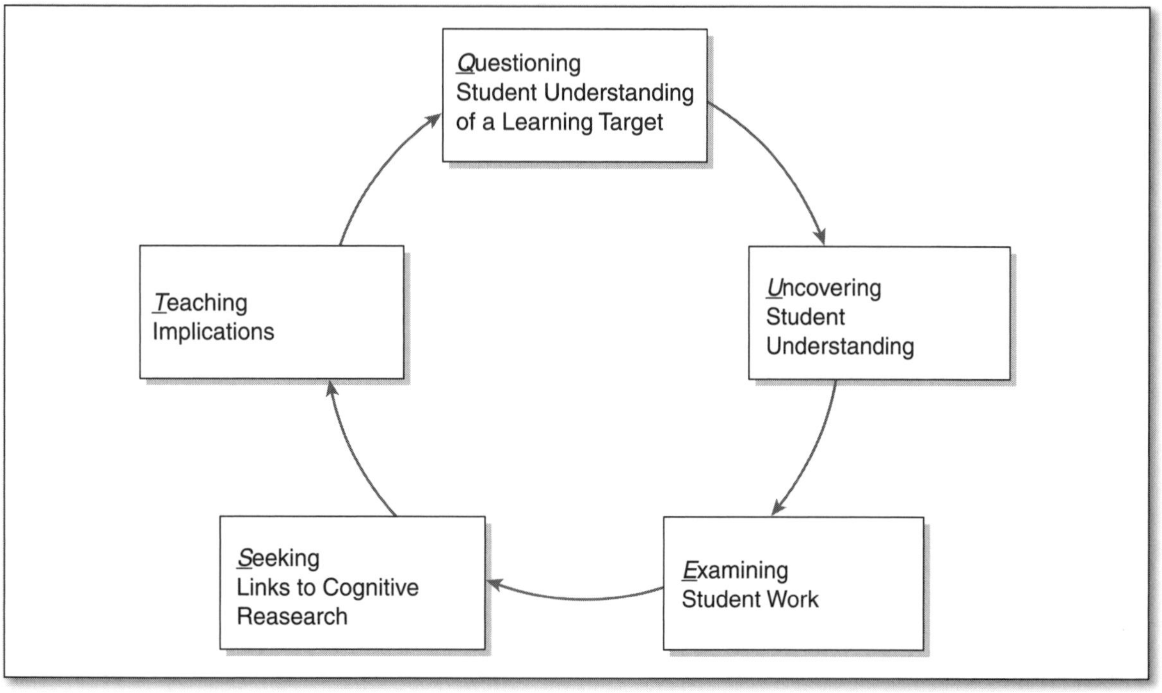

Source: Rose, C., Minton, L., & Arline, C. (2007).

Questioning for Student Understanding

This component helps to focus a teacher on what a particular probe elicits and to provide information on grade-appropriate knowledge. Figure 1.16 shows an example question from the Mathematics Assessment Probe, What's Your Addition Strategy?

Figure 1.16 Sample Question From the What's Your Addition Strategy? Probe

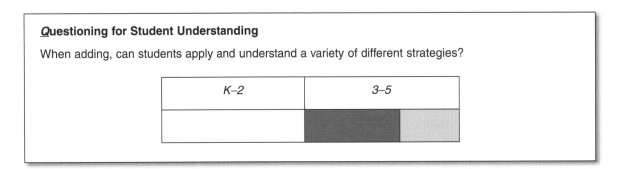

Grade-span bars are provided to indicate the developmentally appropriate level of mathematics as aligned to the NCTM Standards and cognitive research. The dark gray band represents the grade levels where the mathematics required of the probe is aligned to the standards, and the lighter gray band shows grade levels where field testing of the probe has indicated students still have difficulties. The grade spans, although aligned to the standards, should be considered benchmarks as some students at higher grades may have misunderstandings based in understandings from lower grades, while others may be further along the learning progression and need probes designed for older students.

Uncovering Understanding

Figure 1.17 shows an example, Uncovering Understanding, from the Mathematics Assessment Probe, What's Your Addition Strategy?

Following the Teacher Notes and sample student responses, adaptations or variations of the Mathematics Assessment Probe are provided for some of the probes. Variations of the probes provide a different structure (selected response, multiple selections, opposing views, examples/nonexamples, justified list, and strategy harvest) for the question within the same grade span. An adaptation to the probe is similar in content to the original, but the level of mathematics changes for use with a different grade span. When a variation and/or adaptation of the probe is provided, the information is included in the Uncovering Understanding section.

Figure 1.17   Uncovering Understanding

<u>U</u>ncovering Understanding

Addition Strategies: Whole Numbers Content Standard: Number and Operation

Important Note: Prior to giving students the probe, ask them to individually solve the indicated problem. If time allows, ask them to solve the problem in at least two or three different ways.

Variations/Adaptations:
- ○ Addition Strategies: Three-Digit Numbers
- ○ Addition Strategies: One-Digit Numbers

Examining Student Work

This section includes information about the stem, answers, and distracters as related to the research on cognitive learning. Example student responses are given for a selected number of elicited understandings and misunderstandings. The categories, conceptual/procedural and common errors/overgeneralizations, are used where appropriate and are written in italics. Figure 1.18 shows an example of Examining Student Work, from the Mathematics Assessment Probe, What's Your Addition Strategy?

Figure 1.18   Sample of Examining Student Work From the What's Your Addition Strategy? Probe

---

<u>E</u>xamining Student Work

Student answers may reveal *misunderstandings* regarding methods of addition, including a lack of *conceptual understanding* of number properties. Responses also may reveal a common misconception that there is only one correct algorithm for each operation or that, once comfortable with a method, there is no need to understand other methods.

- Sam's method: This method is usually recognized by third-to-fifth-grade students, although in some situations students may not have been introduced to this standard U.S. algorithm. Those who have no experience with the method show lack of *procedural understanding* of the algorithm and typically indicate the method "does not make sense because 1 + 34 + 56 is 91, not 90." Those students who do recognize the algorithm often do not demonstrate place value understanding. (See Student Responses 1 and 2, Chapter 5, page 133.)
- Julie's method: This method is often recognized by students who have experience with multiple algorithms as well as those who were taught only the traditional algorithm. These latter students often apply variations using an expanded notation form of the numbers. (See Student Response 3, Chapter 5, page 133.)
- Pete's method: This strategy is often the least recognized by students in terms of generalizing a method of adding and subtracting like amounts from the numbers to keep a constant total. (See Student Response 4, Chapter 5, page 133.)
- Lisa's method: Students who use this strategy hold the first number constant and break the addend into place value parts, adding on one part at a time. Students who have experience using an open number line are typically able to mathematically explain this method of addition. (See Student Response 5, Chapter 5, page 133.)

Seeking Links to Cognitive Research

This section provides additional information about research that teachers can use for further study of the topic. Figure 1.19 shows an example from the Mathematics Assessment Probe, What's Your Addition Strategy?

Figure 1.19   Seeking Links to Cognitive Research

<u>S</u>eeking Links to Cognitive Research

Student errors when operating on whole numbers suggest students interpret and treat multi digit numbers as single-digit numbers placed adjacent to each other, rather than using place-value meanings for digits in different positions. (AAAS, 1993, p. 358)

The written place-value system is a very efficient system that lets people write very large numbers. Yet it is very abstract and can be misleading: The digits in every place look the same. To understand the meaning of the digits in the various places, children need experience with some kind of *size-quantity supports* (e.g., objects or drawings) that show tens to be collections of 10 ones and show hundreds to be simultaneously 10 tens and 100 ones, and so on. (NCTM, 2003, p. 78)

Students can use roughly three classes of effective methods for multi digit addition and subtraction, although some methods are mixtures. *Counting list methods* are extensions of the single-digit counting methods. Children initially may count large numbers by ones, but these unitary methods are highly inaccurate and are not effective. All children need to be helped as rapidly as possible to develop prerequisites for methods using tens. These methods generalize readily to counting on or up by hundreds but become unwieldy for larger numbers. In *decomposing methods,* children decompose numbers so that they can add or subtract the like units (add tens to tens, ones to ones, hundreds to hundreds, etc.). These methods generalize easily to very large numbers. *Recomposing methods* are like the make-a-ten or doubles methods. The solver changes both numbers by giving some amount of one number to another number (i.e., in adding) or by changing both numbers equivalently to maintain the same difference (i.e., in subtracting). (NCTM, 2003, p. 79)

When students merely memorize procedures, the may fail to understand the deeper ideas. When subtracting, for example, many children subtract the smaller number from the larger in each column, no matter where it is. (National Research Council, 2002b, p. 13)

By the end of the 3–5 grade span, students should be computing fluently with whole numbers. *Computational fluency* refers to having efficient and accurate methods for computing. Students exhibit computational fluency when they demonstrate flexibility in the computational methods they choose, understand and can explain these methods, and produce accurate answers efficiently. The computational methods that a student uses should be based on mathematical ideas that the student understands well, including the structure of the base-ten number system. (NCTM, 2000, p. 152)

Computation skills should be regarded as tools that further understanding, not as a substitute for understanding. (Paulos, 1991, p. 53)

When students merely memorize procedures, they may fail to understand the deeper ideas that could make it easier to remember—and apply—what they learn. Understanding makes it easier to learn skills, while learning procedures can strengthen and develop mathematical understanding. (NRC, 2002, p. 13)

Study results indicate that almost all children can and do invent strategies and that this process of invention (especially when it comes *before* learning standard algorithms) may have multiple advantages. (NCTM, 2002c, p. 93)

Teaching Implications

Being aware of student difficulties and their sources is important, but designing instruction is most critical in helping diminish those difficulties. Although some ideas are included, through "Focus Through Instruction" statements and "Questions to Consider," the authors strongly encourage educators to use the *Curriculum Topic Study (CTS)* (Keeley & Rose, 2006) process to search for additional teaching implications. Each set of Teacher Notes refers the related *CTS* guide for further study, additional references, and a Teacher Sound Bite from a field tester of the probe. Figure 1.20 shows an example from the Mathematics Assessment Probe, What's Your Addition Strategy?

Figure 1.20   Teaching Implications

Focus Through Instruction

- Focusing on understanding multidigit addition methods results in much higher levels of correct use of methods.
- Students need visuals to understand the meanings of hundreds, tens, and ones. These meanings need to be related to the oral and written numerical methods developed in the classroom.
- Number lines and hundreds grids support counting-list methods the most effectively.
- Decomposition methods are facilitated by objects that allow children to physically add or remove different quantities (e.g., base-10 blocks).
- Student's who believe there are several correct methods for adding numbers often show higher engagement levels.
- When children solve multidigit addition and subtraction problems, two types of problem-solving strategies are commonly used: invented strategies and standard algorithms. Although standard algorithms can simplify calculations, the procedures can be used without understanding, and multiple procedural issues can exist.
- Both invented and standard algorithms can be analyzed and compared, helping students understand the nature and properties of the operation, place value concepts for numbers, and characteristics of efficient methods and strategies.

Questions to Consider (when working with students as they develop and/or interpret a variety of algorithms)

- When exploring and/or inventing algorithms, do students consider the generalizability of the method?
- Are students able to decompose and recompose the type of number they are operating with?
- Can students explain why a strategy results in the correct answer?
- When analyzing a strategy or learning a new method, do students focus on properties of numbers and the underlying mathematics rather than just memorizing a step-by-step procedure?
- Do students use a variety of estimation strategies to check the reasonableness of the results?

Teacher Sound Bite

I struggle to know what methods students bring with them each year when transitioning to my fourth-grade class. These strategy probes help me consider my students' level of comfort and experience with a variety of ways to add numbers and whether they have more than just an understanding of the steps a procedure.

(Continued)

Figure 1.20 (Continued)

**Curriculum Topic Study
and Uncovering
Student Thinking**

Place Value

What's Your Addition Strategy? (page 128)

Keeley, P., & Rose, C. (2007). *Mathematics curriculum topic study: Bridging the gap between standards and practice.* Thousand Oaks, CA: Corwin. (Addition and Subtraction, p. 111).

Related Elementary Probes:

Rose, C., & Arline, C. (2009). *Uncovering student thinking in mathematics, grades 6–12: 30 formative assessment probes for the secondary classroom.* Thousand Oaks, CA: Corwin. (Variation: What's Your Addition Strategy? Decimals, p. 82; Fractions, p. 83).

Additional References for Research and Teaching implications

McREL. (2002). *EDThoughts: What we know about mathematics teaching and learning.* Bloomington, IN: Solution Tree. (pp. 82–83).

National Council of Teachers of Mathematics. (2000). *Principles and standards for school mathematics.* Reston, VA: Author. (p. 152).

National Council of Teachers of Mathematics. (2002). Lessons learned from research. Reston, VA: Author. (pp. 93–100).

National Council of Teachers of Mathematics. (2003). *Research companion to principles and standards for school mathematics.* Reston, VA: Author. (pp. 68–84).

National Research Council. (2002). *Helping children learn mathematics.* Washington, DC: National Academy Press. (pp. 11–13).

National Research Council. (2005). *How students learn: Mathematics in the classroom.* Washington, DC: National Academy Press. (pp. 223–231).

Paulos, J. A. (1991). *Beyond numeracy.* New York: Vintage. (pp. 52–55).

A note about the use of interactive technology applets: Some of the concepts elicited by the probes can be addresed through available online resources. Keep in mind that most of these applets were developed as instructional resources or to provide practice and were not developed to address a specific misconception. When searching for available applets that meet students' needs as elicited by a probe, be sure to review the applet carefully to consider the range of examples and nonexamples that can be modeled using the tool. Before using with students, prepare a scaffolded set of questions designed specifically to highlight the misunderstandings elicited by the probe items.

Following is a sample list of sites to look at for freely available interactive applets:

- National Library of Virtual Manipulatives (http://nlvm.usu.edu/en/nav/vlibrary.html)
- NCTM's Illuminations (http://illuminations.nctm.org)
- Educational Development Center (http://maine.edc.org/file.php/1/K6.html)

In addition to the Teacher Notes, a Note Template is included in the back of the book (see Resource: Notes Template: QUEST Cycle). The Note Template provides a structured approach to working through a probe quest. The components of the template are described in Figure 1.21.

Figure 1.21   Notes Template: QUEST Cycle

_Q_uestioning for Student Understanding of a Particular Concept

Considerations: What is the concept you wish to target? Is the concept at grade level or is it a prerequisite?

_____

_____

_____

_U_ncovering Understandings and Misunderstandings Using a Probe

Considerations: How will you collect information from students (paper-pencil responses, interview, student response system, etc.)? What form will you use (one-page probe, card sort, etc.)? Are there adaptations you plan to make? Review the summary of typical student responses. What do you predict to be common understandings and/or misunderstandings for your students?

_____

_____

_____

_E_xamining Student Work

Sort by selected responses then re-sort by trends in thinking
Considerations: What common understanding and misunderstandings were elicited by the probe?

_____

_____

_____

_S_eeking Links to Cognitive Research to Drive Next Steps in Instruction

Considerations: How do these elicited understanding and misunderstandings compare to those listed in the Teacher Notes? Review the bulleted items in the Focus Through Instruction and Questions to Consider to begin planning next steps. What additional sources did you review?

_____

_____

_____

_T_eaching Implications Based on Findings and Determining Impact on Learning by Asking an Additional Question

Considerations: What actions did you take? How did you assess the impact of those actions? What are your next steps?

_____

_____

_____

WHAT MATHEMATICS ASSESSMENT PROBES ARE INCLUDED IN THE BOOK?

Many of the samples included in this book fall under numerical concepts and operations because the cognitive research is abundant in these areas at Grades K –5. The book also includes multiple examples for the following additional content standards: algebra, data analysis, geometry, and measurement. Figure 1.22 provides an "at a glance" look of the grade span and content of the 25 probes with Teacher Notes that are included in Chapters 3 through 6. Grade-span bars are provided to indicate the developmentally appropriate level of mathematics as aligned to NCTM Standards as well as the cognitive research.

Figure 1.22   Table of Probes With Teacher Notes

Grade-Span Bar Key

	Target for Instruction Depending on Local Standards
	Prerequisite Concept and Field Testing Indicate Students May Have Difficulty

Question	Probe	Grade Span					
		K	1	2	3	4	5
Structure of Number: Place Value, Number Charts, and Number Lines							
Do students understand the value of each digit in a double-digit number?	How Many Stars? (page 42)		1–2		3		
Can students translate numbers from verbal to symbolic representation?	What's the Number? (page 49)		1–2		3		
Can students choose all correct values of various digits of a given whole number?	What Is the Value of the Place? (page 54)		1–3			4	
Are students able to estimate a value on a number line given the value of the endpoints?	What Number Is That? (page 58)		1–2		3–4		
Given a portion of the 100s chart, can students correctly fill in the missing values?	Hundred Chart Chunks (page 64)			2–3		4–5	
Can students choose all correct values of various digits of a given decimal?	What Is the Value of the Place? (page 71)				3–4		5
Structure of Number: Parts and Wholes							
Do students understand the meaning of the equal sign?	Equal to 4? (page 78)				2–4		5
Given a set model, are students able to define fractional parts of the whole?	Crayon Count (page 85)				2–4		5
Given a whole, are students able to indentify when $\frac{1}{4}$ of the whole is shaded?	Is $\frac{1}{4}$ of the Whole Shaded? (page 91)				3–4		5
Given an area model, are students able to define fractional parts of the whole?	Granola Bar (page 99)					4–5	
Are students able to choose equivalent forms of a fraction?	Is It Equivalent? (page 104)					4–5	
Do students use the "cancelling of zeros" shortcut appropriately?	Is It Simplified? (page 110)					4–5	

Question	Probe	K	1	2	3	4	5
Structure of Number: Computation and Estimation							
Do students use the structure of ten when combining collections?	How Many Dots? (page 116)	K	1–2				
When considering the whole and two parts, can students identify all possible part-part-whole combinations?	Play Ball (page 122)		1–2		3		
When adding, can students apply and understand a variety of different strategies?	What's Your Addition Strategy? (page 128)			2–3		4–5	
When subtracting, can students apply and understand a variety of different strategies?	What's Your Subtraction Strategy? (page 136)			2–3		4–5	
Are students flexible in using strategies for solving various multiplication problems?	What's Your Multiplication Strategy? (page 144)					4–5	
Are students flexible in using strategies for solving division problems?	What's Your Division Strategy? (page 150)					4–5	
Do students understand there are multiple methods of estimating the sum of three 2-digit numbers?	Is It an Estimate? (page 155)					4–5	
Can students use estimation to choose the closest benchmark to an addition problem involving fractions?	What Is Your Estimate? (page 160)					4–5	
Measurement and Geometry							
Do students understand the properties and characteristics of quadrilaterals?	Quadrilaterals (page 166)				3–4		5
Are students able to determine area without the typical length by width labelling?	What's the Area? (page 174)				3–5		
Do students pay attention to starting point when measuring with nonstandard units?	What's the Measure? (page 180)				3–5		
Data							
Are students able to choose the correct graphical representation when given a mathematical situation?	Graph Choices (page 187)				3–5		
Do students understand ways the median is affected by changes to a data set?	The Median (page 194)				3–5		

In addition to the 25 probes with teacher notes, many variations to the probes are provided. Although some of the variations provide a different structure option, others are created to target a different grade span by addressing a foundational concept to the original probe. For these foundational variations to the probe, information from the Teacher Notes, although helpful, is not intended to align directly. Therefore we strongly suggest you use the QUEST cycle template to create your own version of teacher notes. The following table provides a grade-level view of the probes and variations that are included in this resource.

Figure 1.23   Table of Probes by Targeted Grade Span

Grade K		
Probe	*Grade Span*	*Chapter/Page*
Variation: How Many Counters?	K	Chapter 5, page 121
How Many Dots?	K	Chapter 5, page 116
Variation: What's the Measure?	K–1	Chapter 6, page 186
Variation: Is It a Circle?	K–1	Chapter 6, page 173
Grade 1		
Probe	*Grade Span*	*Page*
What Is the Value of the Place?	1–2	Chapter 3, page 54
What's the Number?	1–2	Chapter 3, page 49
What Number Is That?	1–2	Chapter 3, page 58
How Many Stars?	1–2	Chapter 3, page 42
Variation: How Many Counters?	1	Chapter 5, page 121
Play Ball	1–2	Chapter 5, page 122
Variation: What's the Measure?	K–1	Chapter 6, page 186
Variation: Is It a Circle?	K–1	Chapter 6, page 173
Grade 2		
Probe	*Grade Span*	*Page*
What's the Number?	1–2	Chapter 3, page 49
What Is the Value of the Place?	2–3	Chapter 3, page 54
What Number Is That?	1–2	Chapter 3, page 58
Hundred Chart Chunks	2–3	Chapter 3, page 64
Equal to 4?	2–4	Chapter 4, page 78
Crayon Count	2–4	Chapter 4, page 85
Play Ball	1–2	Chapter 5, page 122
What's Your Addition Strategy?	2–3	Chapter 5, page 128
What's Your Subtraction Strategy?	2–3	Chapter 5, page 136
Variation: Is It a Polygon?	2–3	Chapter 6, page 172
Grade 3		
Probe	*Grade Span*	*Page*
Hundred Chart Chunks	2–3	Chapter 3, page 64
What is the Value of the Place?	2–3	Chapter 3, page 54
What Is the Value of the Digit?	3–5	Chapter 3, page 71
Equal to 4?	2–4	Chapter 4, page 78
Crayon Count	2–3	Chapter 4, page 85
Is $\frac{1}{4}$ of the Whole Shaded?	3–4	Chapter 4, page 91

Probe	Grade Span	Page
What's Your Addition Strategy?	2–3	Chapter 5, page 128
What's Your Subtraction Strategy?	2–3	Chapter 5, page 136
Quadrilaterals	3–4	Chapter 6, page 166
Variation: What's the Measure?	3–5	Chapter 6, page 186
What's the Area?	3–5	Chapter 6, page 174
Graph Choices	3–5	Chapter 6, page 187
The Median	3–5	Chapter 6, page 194
Variation: Is It a Polygon?	2–3	Chapter 6, page 172

Grade 4

Probe	Grade Span	Page
What Is the Value of the Digit?	3–5	Chapter 3, page 71
Equal to 4?	2–4	Chapter 4, page 78
Is $\frac{1}{4}$ of the Whole Shaded?	3–4	Chapter 4, page 91
Granola Bar	4–5	Chapter 4, page 99
Is It Equivalent?	4–5	Chapter 4, page 104
Is It Simplified?	4–5	Chapter 4, page 110
Is It an Estimate?	4–5	Chapter 5, page 155
What's Your Multiplication Strategy?	4–5	Chapter 5, page 144
What's Your Division Strategy?	4–5	Chapter 5, page 150
What Is Your Estimate?	4–5	Chapter 5, page 160
Quadrilaterals	3–4	Chapter 6, page 166
What's the Measure?	3–5	Chapter 6, page 180
Variation: Length of a Line	4–5	Chapter 6, page 185
What's the Area?	3–5	Chapter 6, page 174
Graph Choices	3–5	Chapter 6, page 187
The Median	3–5	Chapter 6, page 194

Grade 5

Probe	Grade Span	Page
What Is the Value of the Digit?	3–5	Chapter 3, page 71
Granola Bar	4–5	Chapter 4, page 99
Is It Equivalent?	4–5	Chapter 4, page 104
Is It Simplified?	4–5	Chapter 4, page 110
Is It an Estimate?	4–5	Chapter 5, page 155
What's Your Multiplication Strategy?	4–5	Chapter 5, page 144
What's Your Division Strategy?	4–5	Chapter 5, page 150
What Is Your Estimate?	4–5	Chapter 5, page 160
What's the Measure?	3–5	Chapter 6, page 180
Variation: Length of a Line	4–5	Chapter 6, page 185
What's the Area?	3–5	Chapter 6, page 174
Graph Choices	3–5	Chapter 6, page 187
The Median	3–5	Chapter 6, page 194

Instructional Implications

Assessment for Learning is part of everyday practice by students, teachers and peers that seeks, reflects upon and responds to information from dialogue, demonstration and observation in ways that enhance ongoing learning.

—Assessment Reform Group (2002)

Early research by Reynolds (1993) and Shulman (1987) determined that teaching and learning were both influenced by knowledge, implementation, and analysis as a cyclic model of teacher practice. Use of the Mathematics Assessment Probes provides the opportunity to replicate this cycle to inform instruction with a solid research base to support increased content knowledge, a specific structure and purpose for implementation, and two tiers of responses to uncover student thinking. Effective teaching demands analysis of both student performance and one's own teaching performance. A teacher's knowledge base is enhanced when a teacher reflects on his or her practice by listening to students, analyzing student work, and reviewing student responses to class activities

Mathematics Assessment Probes represent an approach to diagnostic assessment. They can be used for formative assessment purposes if the information about students' understandings and misunderstandings is used to focus instruction. Purposes for using the Mathematics Assessment Probes included in this chapter are

- differentiating instruction;
- assessing point of entry;
- analyzing trends in thinking;
- giving student interviews;
- promoting math conversations;

- developing vocabulary;
- allowing for individual think time;
- improving communication and process skills;
- assessing effectiveness of instructional activities and materials; and
- building capacity across grade levels and spans.

Each of these contexts is briefly described below, and in some cases, Images From Practice are used to highlight strategies within the contexts. The Images From Practice provide a window into the classroom of a teacher who uses the Mathematics Assessment Probes.

DIFFERENTIATING INSTRUCTION

Not all students are alike. On the basis of this knowledge, differentiated instruction applies an approach to teaching and learning that gives students multiple options for taking in information and making sense of ideas. The theory of differentiated instruction is based on the premise that instructional approaches should vary and be adapted in relation to individual and diverse students in classrooms (Tomlinson, 2001). The model of differentiated instruction requires teachers to be flexible and adjust the curriculum and presentation of information to learners. (Berkas & Pattison, 2008, para. 3)

A differentiated classroom offers a variety of learning options designed to tap into different readiness levels, interests, and learning profiles. In a differentiated class, the teacher uses (1) a variety of ways for students to explore curriculum content, (2) a variety of sense-making activities or processes through which students can come to understand and "own" information and ideas, and (3) a variety of options through which students can demonstrate or exhibit what they have learned. A class is not differentiated when assignments are the same for all learners and the adjustments consist of varying the level of difficulty of questions for certain students, grading some students harder than others, or letting students who finish early play games for enrichment. It is not appropriate to have more advanced learners do extra math problems, do extra book reports, or—after completing their "regular" work—be given extension assignments. Asking students to do more of what they already know is hollow. Asking them to do "the regular work, plus" inevitably seems punitive to them (Tomlinson, 1995).

ASSESSING POINT OF ENTRY

A major concern from both a philosophical and pedagogical perspective is that, because children develop and learn at individually different rates, no one set of age-related goals can be applied to all children. A specific learning time line may create inaccurate judgments and

categorizations of individual children, limit the curriculum to those out-
comes, lead to inappropriate teaching of narrowly defined skills, and
limit the development of the "whole child." (Bredekamp, 2004, p. 167)

Assessing prior knowledge is a key first step in using assessment to inform
teaching. Often, assumptions are made about what students do or do not understand. Assumptions about lack of readiness may be based on teachers' experiences with students' lack of understanding in previous years. Assumptions that
students are ready are often based on the fact that the students have studied the
materials before (Stepans, Schmidt, Welsh, Reins, & Saigo, 2005).

Much has been learned regarding young children's innate understanding
about number. Young children possess an informal knowledge of mathematics
that is surprisingly broad, complex, and sophisticated. Lindquist and Joyner
(2004) discovered that most beginning kindergarteners show a surprisingly
high entry level of mathematical skills. Mathematical knowledge begins during
infancy and undergoes extensive development over the first five years of life. It
is as natural for young children to think mathematically as it is for them to use
language because "humans are born with a fundamental sense of quantity, as
well as spatial sense and a propensity to search for patterns" (Geary, 1994,
p. 46). Although young children possess rich experiential knowledge, they do
not have equal opportunities to bring this knowledge to an explicit level of
awareness. These opportunities are essential for children to be able to connect
their informal mathematical experiences to school mathematical experiences.

It is essential for our youngest students that we are able to assess points of
entry for them that are based on data and not assumptions. Just as prevalent as
learning targets that are too difficult for students is the wasting of valuable
classroom time by incorporating activities below the instructional level of the
students. Knowing the "instructional level" for a student in mathematics is
essential in maximizing instructional time and experiences. As in reading, both
instructional and independent levels should be known and nurtured through
thoughtful scaffolding of meaningful mathematics for all students. The
Mathematics Assessment Probes can be given prior to a specific unit of investigation to gauge the starting point and allow teachers to make decisions based
on evidence rather than assumption.

ANALYZING TRENDS IN THINKING

"Compiling an inventory for a set of papers can provide a sense of the class's
progress and thus inform decisions about how to differentiate instruction"
(Burns, 2005, p. 29).

In her recent article, "Looking at How Students Reason," Marilyn Burns
(2005) describes a process for taking a classroom inventory:

After asking a class of 27 fifth graders to circle the larger fraction ($\frac{2}{3}$ or
$\frac{3}{4}$) and explain their reasoning, I reviewed their papers and listed the
strategies they used. Their strategies included drawing pictures (either

circles or rectangles); changing to fractions with common denominators ($\frac{8}{12}$ and $\frac{9}{12}$); seeing which fraction was closer to 1 ($\frac{2}{3}$ is $\frac{1}{3}$ away, but $\frac{3}{4}$ is only $\frac{1}{4}$ away); and relating the fractions to money ($\frac{2}{3}$ of $1.00 is about 66 cents, whereas $\frac{3}{4}$ of $1.00 is 75 cents). Four of the students were unable to compare the two fractions correctly. I now had direction for future lessons that would provide interventions for the struggling students and give all the students opportunities to learn different strategies from one another. (p. 29)

Developing this "sense of the class" allows for instructional decision making. The probes can be used for this purpose by categorizing student responses and asking the following questions:

- What are the primary methods students used for solving this problem?
- How often do the primary methods result in the correct response?
- Which of the methods are generalizable?
- What student methods are considered outliers?
- Which of the methods are more efficient?
- Which methods are explored within the math curriculum materials?
- Based on the sense of the class, what instructional strategies are effective for this particular learning target?

The last question is an important one because students' preconceptions must be addressed explicitly in order for them to change their existing understanding and beliefs. As students begin to grab onto what makes sense to them, misconceptions and partial understandings can become very ingrained if not surfaced and if students are not given an opportunity to work to "undo" incorrect thinking. Students will hold onto faulty ideas and procedures and new ideas will be layered on top of them, making it difficult to uncover the core idea as it is embedded in student understanding. Identifying misunderstanding, misconceptions or partial understandings is necessary to be able to "undo" incorrect thinking before it is ingrained and therefore embedded in other ideas and beliefs that make isolating and modifying the incorrect ideas more difficult. If students' initial ideas are ignored, the understanding that they develop can be very different from what the teacher intends. (Stepans et al., 2005, p. 35)

GIVING STUDENT INTERVIEWS

"Interviewing provides the opportunity to talk with students—that is, to hear their explanations and to pose follow-up questions that probe the rationale behind their beliefs" (Stepans et al., 2005, p. 36).

Giving student interviews is an important and useful strategy for several purposes. The main purpose for interviewing a student is to probe that student's mathematical thinking. By interviewing a variety of students in a class, [teachers] can get a better sense of the range of thinking in that class. Interviewing takes time, but the potential payoff is great for helping make sense of students' responses to questions. (Bright & Joyner, 2004, p. 184)

The following Image From Practice highlights how a teacher used the probe How Many Dots? as a tool to interview students to uncover what they understood about the structure of ten.

Image From Practice: How Many Dots?

I watched my first graders count, read, identify, and join two-digit numbers through 50, so I believed that they understood the "ten-ness" of our number system. We use Popsicle sticks during calendar time and dimes and pennies for days in school, so I was very surprised, when I used the How Many Dots? Probe, how many of my students counted by ones. I have used many models and representations and mistaken students' rote responses to suggest that they understood the structure of ten and, what is more important, that they were using it to add. As a result of using the probe and examining my previous materials and instruction, I am now adding reasoning questions to my math routines and making explicit connections to ensure that I am providing students with the opportunity to respond at both the conceptual and procedural levels during all of our classroom math routines.

PROMOTING MATH CONVERSATIONS

Like a piece of music, the classroom discourse has themes that pull together to create a whole that has meaning. The teacher has a role in orchestrating the oral and written discourse in ways that contribute to students' understanding of mathematics. (National Council of Teachers of Mathematics [NCTM], 2003, p. 46)

The Mathematics Assessment Probes can be used to begin classroom conversations based on the results of using a probe, or they can be the impetus for the conversation. Building a school and class culture that fosters talk about mathematical thinking is important in setting an expectation that "math talk" is essential. As the quote above suggests, discourse is necessary in order to support student reasoning and justification. Sharing ideas by making them accessible and relevant allows teachers to integrate specific ideas, skills, and strategies into their daily instruction to further understanding for all students who may be at different points along the way to grasping a significant math idea.

The following Image From Practice highlights how a teacher used the results from the probe Play Ball to address a lack of flexibility in using part-part-whole number concepts she had previously assumed her students understood. This idea is necessary for students to develop in preparation for computation and algebraic thinking.

Image From Practice: Play Ball

After giving the Play Ball probe, it became very clear to me that my students were able to recall number fact combinations, especially the doubles, with great enthusiasm and accuracy when asked to respond to a particular prompt. What was missing from my instruction, as indicated by the probe, was providing students with the opportunity to consider that multiple combinations could be used within a context to answer a question. Most of my students answered 4 soccer balls and 4 baseballs with the reasoning that 4 + 4 = 8. It did not occur to them that there were other possibilities for the combinations of balls. As a result, I am adding a scenario similar to Play Ball each week during morning messages to provide my students with the opportunity to move past their initial response and determine if there are other possibilities. Sharing their thinking allows everyone an access point and encourages them to think more deeply to decide if they have found all the possible combinations. This is an important idea as students begin to use larger numbers and think about how to compose and decompose them efficiently. If they develop only one combination for a particular number, they will have great difficulty generalizing their knowledge to larger numbers, which is likely to impede their computational fluency.

Using purposeful math conversations with students allows them to become aware of their own thinking and therefore enables them to modify their thinking if they need to. This process allows students to practice metacognition strategies and further develop communication skills that put focus on both the answer and the process. When asked to engage in similar problems, students will be better able to apply their new or modified ideas, and they will also be able to discuss how they thought about the problem with peers.

DEVELOPING VOCABULARY

Children learn vocabulary primarily indirectly through their conversations with others and the books and programs they're exposed to. However, many words used in mathematics may not come up in everyday contexts—and if they do, they may mean something totally different—so, math vocabulary needs to be explicitly taught. (Minton, 2007, p. 40)

Students' use of math terms is directly related to their experiences. Lack of exposure to math situations and opportunities to develop a correct mathematical vocabulary can deprive students of the language of math. The language of math is specific and uses not only words to denote meaning but also symbolic notation. Symbols enable mathematical ideas to be expressed in precise ways that reflect quantitative relationships. Misunderstanding of the meaning of symbols or misconception of how to use symbols in math can impact understanding. Words can take on different meanings in different contexts, as can symbols. Using the Mathematics Assessment Probes can be a tool in helping students to build strong math vocabularies that enable them to make connections between concepts. A probe structure that is useful in helping to build vocabulary is the examples/nonexamples format.

The examples and nonexamples format of some of the Mathematics Assessment Probes provide an important tool for assessing and building conceptual knowledge by eliciting prior knowledge from students. Small-group and whole-class discussions focused on choices lead students to justify their categorizations. The following Image From Practice highlights how a teacher used the probe What Is the Value of the Place? to enhance students' understanding of the concept of place value.

Image From Practice: What Is the Value of the Place?

I decided to use the probe What Is the Value of the Place? before I began the first unit in my third-grade math program. My students seemed able to identify three-digit numbers with ease, yet their flexibility in composing and decomposing numbers by place value was inconsistent. I had hoped that by using the probe I would get a sense of how my students were thinking about the value of the places in a three-digit number and to look more closely at what they understood about our base-10 number system.

By looking at my students' thinking on the probe, I could see a significant number of students treated digits as their face value, rather than their place value, and for the most part thought in terms of ones only. Many did not think of the 74 in 749 as 74 tens. They were responding to the numbers as a whole and not using place value to see the numbers in various forms of component parts. I used the information from the probe to be explicit and purposeful with the mathematical tools (manipulatives) and activities and games of the unit to develop a shared definition of place value and had my students revisit different-sized numbers through our morning messages to reinforce the concept of place value.

ALLOWING FOR INDIVIDUAL THINK TIME

If questions are vehicles for thought, then the questioning process determines who will go along for the ride. Teacher-questioning behaviors affect which students learn how much . . . Another way teachers influence student learning via questioning is through the use of wait time. The tendency to wait (or not) for a student response has been found to vary among teachers and influence student engagement and subsequent learning. (Walsh & Sattes, 2005, p. 73)

Eliciting student ideas using a Mathematics Assessment Probe allows for individual students to express their initial thinking—the first phase of eliciting student thinking—without the influence of other students' ideas. Although student conversation about the learning target addressed in the probe is important, the conversation does not necessarily need to occur by having students discuss the probe itself. Rather, the student conversation takes place during activities specifically chosen to meet the needs of the students, based on evidence of understanding uncovered by the probe. The following excerpt from the article, "Classroom Assessment: Minute by Minute, Day by Day" (Leahy, Lyon, Thompson, & Wiliam, 2005) provides additional ideas for allowing for individual think time. The strategies described below can be incorporated while using a Mathematics Assessment Probe.

Teachers can also use questions to check on student understanding before continuing the lesson. We call this a "hinge point" in the lesson because the lesson can go in different directions, depending on student responses. By explicitly integrating these hinge points into instruction, teachers can make their teaching more responsive to their students' needs in real time. However, no matter how good the hinge-point question, the traditional model of classroom questioning presents two additional problems. The first is lack of engagement. If the classroom rule dictates that students raise their hands to answer questions, then students can disengage from the classroom by keeping their hands down.

The second problem with traditional questioning is that the teacher gets to hear only one student's thinking. To gauge the understanding of the whole class, the teacher needs to get responses from all the students in real time. One way to do this is to have all students write their answers on individual dry-erase boards, which they hold up at the teacher's request. The teacher can then scan responses for novel solutions as well as misconceptions.

Another approach is to give each student a set of four cards labeled A, B, C, and D, and ask the question in multiple-choice format. If the question is well designed, the teacher can quickly judge the different levels of understanding in the class. If all students answer correctly, the teacher can move on. If no one answers correctly, the teacher might choose to re-teach the concept. If some students answer correctly and some answer incorrectly, the teacher can use that knowledge to engineer a whole-class discussion on the concept or match up the students for peer teaching. Hinge-point questions provide a window into students' thinking and, at the same time, give the teacher some ideas about how to take the students' learning forward. (p. 22)

IMPROVING COMMUNICATION AND PROCESS SKILLS

"Examining and discussing both exemplary and problematic pieces of mathematics writing can be beneficial at all levels" (NCTM, 2000, p. 62).

Although the primary purpose of a Mathematics Assessment Probe is to elicit understandings, partial understandings, and misunderstandings, a secondary benefit is the improvement of students' written and verbal communication skills. After instruction and discussion of the underlying mathematics, individual student responses to a probe can be assessed on the degree to which the response is complete, accurate, and reasonable given the problem. This practice provides a benefit for students to be better prepared for local and state assessments that utilize an open-response format and evaluate student responses according to accuracy and communication of process.

By creating exemplar sets of responses that meet predetermined criteria, students can see how to become more effective communicators and work towards improving their own communication and process skills.

The following Image From Practice highlights how a teacher used the probe Is It Equal to 4? to improve students' methods of justification.

Image From Practice: Is It Equal to 4?

Many of my students were able to solve missing addend or change unknown problem types with ease. They seemed to understand the balance that was embedded in an equation; however, their justification or reasoning of why their answers were correct rested on their fact knowledge and not their understanding of equality as an algebraic idea. I used the results to address the concept of equality in my instruction and also used the results as a I collected the completed probes and organized them into groups of all correct responses with clear communication of true math statements, mixed responses with unclear communication, and incorrect responses and unclear communication. I then created card sets with the same collection in each set of responses of answers and communication. I gave a set of cards to groups of four students and gave them directions to match the answers with a communication response. The goal was to have students talk about what made it easier to match answers with their reasoning, which became our criteria for *a good math response.*

ASSESSING EFFECTIVENESS OF INSTRUCTIONAL ACTIVITIES AND MATERIALS

Mathematical literacy emerges from, among other foundational understandings, a mature sense of number that includes an understanding of place value and comfort with estimating; a data sense that recognizes outliers and misinterpretation of data; a spatial sense that links two- and three-dimensional objects; and a symbol sense that results in algebraic representations that enable generalizations and predictions. Rather than long lists of skills, these concepts must be the foundation of any set of national mathematics curriculum materials. (NCTM, 2003, pp. 31)

Using probes in a pre- and postassessment format allows for evaluation of curriculum and instruction. When implementing a new or revised set of activities, evaluating their impact on student learning is an important component of analyzing the effectiveness of the activities and making further revisions. In using Mathematics Assessment Probes for this purpose, it is important to consider the conceptual understanding or mathematical "big ideas" that are addressed within the activities and that are aligned to local and state learning targets. The probes can be used to assess how well student understanding is predicated on program experiences or built on conceptual understanding that transfers easily to other contexts.

The following Image From Practice highlights how a teacher used the probe Hundred Chart Chunks before and after implementing an instructional unit to assess an increase in students' understanding of the structure of our number system.

Image From Practice: Hundred Chart Chunks

My students have been using the Hundred Chart since Grade 1. They have had many opportunities to use it for finding patterns, playing games, computing, and more. They have copies in their hardcover reference books, we have a large 200 chart on our wall, and our math program refers to it often. I wanted to know if my students could "let go" of the chart as I questioned their conceptual understanding based on how often students appeared to rely on the chart to solve computation problems. I was amazed by the difficulty my students had in completing the chunks from a 100s chart. Without the whole chart visible, students completed the blank spaces with either consecutive numbers or made up a pattern that they considered completed the chunk. What was more interesting to me was that, although my students have been circling, coloring, and reciting the patterns of the horizontal and vertical number grid since Grade 1, they did not appear to understand how the chart is set up. This was a huge eye opener for me and made me reconsider the models used in my program to ensure that my use of the model is consistent with the intent of the mathematics that underlie the model. Given this information, I have transformed my instructional use of the 100s chart, as I am more focused on moving students away from the chart in ways that help them to build a mental model of the base-10 system that works for them.

BUILDING CAPACITY ACROSS GRADE LEVELS AND SPANS

"Every shared vision effort needs, not just a broad vision, but specific realizable goals—milestones we expect to reach before too long. Goals represent what people commit themselves to do" (Senge, Kleiner, Roberts, Smith, & Ross, 1994, p. 121).

Another important opportunity provided by using the probes is that of examining student work with other educators:

> The most important aspect of this strategy is that teachers have access to, and then develop for themselves the ability to understand, the content students are struggling with and ways that they, the teachers, can help. Pedagogical content knowledge—that special province of excellent teachers—is absolutely necessary for teachers to maximize their learning as they examine and discuss what students demonstrate they know and do not know. (Loucks-Horsley, Love, Stiles, Mundry, & Hewson, 2003, p. 183)

By providing a link to *Mathematics Curriculum Topic Study* (Keeley & Rose, 2007) as well as to resources with additional research and instructional implications specific to the ideas of the probe, Mathematics Assessment Probes provide a means for a collaborative approach to examining student thinking and planning for improving instruction. The engine of improvement, growth, and renewal in a professional learning community is collective inquiry. The people in such a school are relentless in questioning the status quo, seeking new methods, testing those methods, and then reflecting on the results (DuFour, DuFour, Eaker, Many, 2006, pp. 68)

The following Image From Practice highlights how a math specialist uses probes to address typical student misunderstandings with other mathematics teachers in the district.

Image From Practice: Using Probes Across the District

As a K–5 math specialist, I have found that using the probes with the teachers in my district has been very effective in transforming our school culture to include talking about mathematics. By using student work from our kids, we remove the "that's not what our kids would do" response. By eliminating student names on probes, all student work becomes part of the whole, and therefore, we look for trends in thinking across a grade level. Then, we take the findings and make decisions about what we think the results mean for us. Using the teacher notes provides the next level of support by providing teachers with research findings and, more importantly, suggestions for instruction that address the math ideas—regardless of a specific math program—that are easily incorporated into math instruction. My teachers have commented on how much they appreciate that we can focus on student learning and impact instruction in a process that is easily incorporated into our school day. One teacher, while taking a graduate course, was required to complete an action research project. She chose to research whether students who did not meet standards on the state test had more misconceptions then those who met the standards. All of the Grades 3 to 5 teachers helped her gather the data needed to complete her project, and the results were very informative to all involved. We did find a higher percentage of students in the "does not meet" category, but the gap wasn't as large as we had predicted.

SUMMARY

In *How Students Learn: Mathematics in the Classroom,* the National Research Council (2005) describes a use of assessment as follows: "Assessments are a central feature of both a learner-centered and a knowledge-centered classroom. They permit the teacher to grasp students' preconceptions, which is critical to working with and building on those notions" (p. 16).

The purposes this chapter described for using Mathematics Assessment Probes represent the multiple ways the probes can support educators in "engaging students' preconceptions and building on existing knowledge" as well as developing an "assessment-centered classroom environment" (National Research Council, 2005, p 12).

3

Structure of Number Probes

Place Value, Number Charts, and Number Lines

Figure 3.1 Chapter 3 Probes

Grade-Span Bar Key

	Target for Instruction Depending on Local Standards
	Prerequisite Concept and Field Testing Indicate Students May Have Difficulty

Question	Probe	Grade Span					
Structure of Number: Place Value, Number Charts, and Number Lines		K	1	2	3	4	5
Do students understand the value of each digit in a double-digit number?	How Many Stars? (page 42)		1–2		3		
Can students translate numbers from verbal to symbolic representation?	What's the Number? (page 49)		1–2		3		
Can students choose all correct values of various digits of a given whole number?	What Is the Value of the Place? (page 54)		1–3			4	
Are students able to estimate a value on a number line given the values of the endpoints?	What Number Is That? (page 58)		1–2		3–4		
Given a portion of the 100 chart, can students correctly fill in the missing values?	Hundred Chart Chunks (page 64)			2–3		4–5	
Can students choose all correct values of various digits of a given decimal?	What Is the Value of the Digit? (page 71)				3–4		5

PROBE 1: HOW MANY STARS?

(Student Interview Task)

How Many Stars Are There?

Student writes answer on line.

How many stars are there?

_____ ☆

- -

Follow the directions below once student writes response.

Teacher *points* to the digit in the ones place of the student response and says,

"Can you use this red pencil to circle how many stars this number means from your answer?"

Next, *point* to the digit in the tens place of the student response and say,

"Now use this blue pencil to circle how many stars this number means from your answer."

TEACHER NOTES: HOW MANY STARS?

*Q*uestioning for Student Understanding

Do students understand the value of a digit as it moves from a single to a double-digit number?

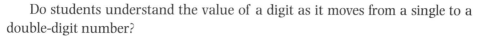

K–2	3–5

*U*ncovering Understanding

Place Value: Whole Numbers Content Standard: Number and Operation

Note: This probe is intended as an interview with one student at a time.

Variation/Adaptation:

- How Many Counters: Hands-On

*E*xamining Student Work

Students answers may reveal *misunderstandings* regarding the value of the *place* rather than the *face* value of the digit. A common misconception may be revealed as students count the stars and then account for the digits without using all of the stars they have just counted in the set.

- Choice A: Student circles 8 stars with the red pencil and 2 stars with the blue pencil. Students who answer this way likely see each digit as a separate quantity with no connection to their place value. (Note: There are a variety of ways students can show what each digit means in their responses. Some may be systematic in their approach to identifying each group and some more random with the overlying big idea of including *all* stars in their final answer as significant.)
- Choice B: Student circles 8 stars with the red pencil and 20 stars with the blue pencil. Students who answer this way recognize that all of the stars need to be included to show 28 and therefore understand the 2 to be a value of 20.

*S*eeking Links to Cognitive Research

During preschool and elementary-school years, children develop meanings for number words in which sequence, count, and cardinal meanings of number words become increasingly integrated. (American Association for the Advancement of Science [AAAS], 1993, p, 350)

Children should learn that the last number named represents the last object as well as the total number of objects in the collection. (National Council of Teachers of Mathematics [NCTM], 2000, p. 79)

Concrete models can help students represent numbers and develop number sense; they can also help bring meaning to students' use of written symbols and can be useful in building place-value concepts. (NCTM, 2000, p. 80)

Teachers should try to uncover students' thinking as they work with concrete materials by asking questions that elicit students' thinking and reasoning. In this way, teachers can watch for students' misconceptions, such as interpreting the 2 tens and 3 ones merely as five objects. (NCTM, 2000, p. 80)

Students also develop understanding of place value through the strategies they invent to compute. (Fuson et al., 1997, p. 169)

Thus, it is not necessary to wait for students to fully develop place-value understandings before giving them opportunities to solve problems with two- and three-digit numbers. (NCTM, 2000, p. 81)

Children develop an understanding of the base-10 numeration system and place-value concepts (at least to 1,000). Their understanding of base-10 numeration includes ideas of counting in units and multiples of hundreds, tens, and ones; as well as a grasp of number relationships, which they demonstrate in a variety of ways, including comparing and ordering numbers. They understand multidigit numbers in terms of place value, recognizing that place-value notation is a shorthand for the sums of multiples of powers of 10 (e.g., 853 as 8 hundreds + 5 tens + 3 ones). (NCTM, 2006, p. 14)

A key idea is that a number can be decomposed and thought about in many ways. (NCTM, 2003, p. 33)

Making a transition from viewing "ten" as simply the accumulation of 10 ones to seeing it both as 10 ones *and* as 1 ten is an important first step for students toward understanding the structure of the base-10 number. (NCTM, 2003, p. 33)

Number sense develops as students understand the size of numbers, develop multiple ways of thinking about and representing numbers, use numbers as referents, and develop accurate perceptions about the effects of operations on numbers.

Concrete models can help students represent numbers and develop number sense; they can also help bring meaning to students' use of written symbols and can be useful in building place-value concepts. (NCTM, 2000, p. 80)

It is essential that students develop a solid understanding of the base-10 numeration system and place-value concepts by the end of grade two. (NCTM, 2000, p. 81)

Teaching Implications

To support a deeper understanding for students in the primary grades in regards to place value, the following are instructional ideas and questions to consider in conjunction with the research.

Focus Through Instruction

- Focus on uncovering students' thinking as they work with concrete materials by asking questions that elicit students' thinking and reasoning.
- Students should recognize that the word *ten* can represent 1 unit or 10 single units.

- Base-10 number knowledge results from an ability to count, to make groupings, and to understand place value at a deeper level than simply naming places.
- Notice what students do computationally that reveals what they understand about place value.
- Choose a variety of numbers and problem types to provide students with opportunity to use their knowledge of place value to develop a variety of strategies.
- Emphasize place-value ideas by asking questions that support the base-10 number system (e.g, "What is ten more? Ten less?").
- Use of calculators can develop and reinforce place-value concepts.

Questions to Consider (when working with students as they develop an understanding of place value)

- When composing and decomposing numbers, do students refer to the quantity of the digit based on their position in the number?
- Do students recognize the relationship between a quantity and its number name?
- Do student strategies vary based on what they understand about the numbers involved in a problem, or do they apply the same strategy regardless of the numbers?
- Are students using number names appropriately when computing?

Teacher Sound Bite

I assumed that because my first graders could count, read, and identify two-digit numbers through 20 that they understood "ten-ness." We do a lot of work with tens as they relate to the teens and 20, yet their understanding of accounting for all of the stars within the 28 they counted was relatively weak. I was so surprised at the number of students who noted the 20 as two stars.

Additional References for Research and Teaching Implications

Burns, M. (2000). *About teaching mathematics: A K–8 resource.* Portsmouth, NH: Heinemann. (p. 173).

National Council of Teachers of Mathematics. (1993). *Research ideas for the classroom: Early childhood mathematics.* Reston, VA: Author. (pp. 53–68).

National Council of Teachers of Mathematics. (2000). *Principles and standards for school mathematics.* Reston, VA: Author. (p. 80–84).

National Council of Teachers of Mathematics. (2003). *Research companion to principles and standards for school mathematics.* Reston, VA: Author. (pp. 33–42).

National Research Council. (2001). *Adding it up: Helping children learn mathematics.* Washington, DC: National Academy Press. (pp. 96–99).

Taylor-Cox, J. (2008). *Differentiating in number and operations.* Portsmouth, NH: Heinemann. (pp. 12–25).

Curriculum Topic Study and Uncovering Student Thinking

How Many Stars?

Keeley, P., & Rose, C. (2006). *Mathematics curriculum topic study: Bridging the gap between standards and practice.* Thousand Oaks, CA: Corwin. (Number Sense, p. 127; Numbers and Number Systems, p. 128; Place Value, p. 130).

Related Probes: What's the Number? (page 49)

Rose, C., Minton, L., & Arline, C. (2007). *Uncovering student thinking in mathematics: 25 formative assessment probes.* Thousand Oaks, CA: Corwin. (Building Numbers, p. 38).

Student Responses to How Many Stars?

Sample Responses: Choice A

Student 1

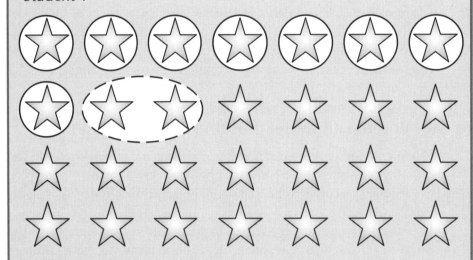

Student writes answer:

How many stars are there?

②⑧ ☆

Student 2

Student writes answer:

How many stars are there?

<u>2 8</u> ☆

Sample Responses: Choice B

Student 3

Student 4

Student writes answer:

How many stars are there?

②⑧ ☆

PROBE 1A: VARIATION: HOW MANY COUNTERS?

(Student Interview Task)

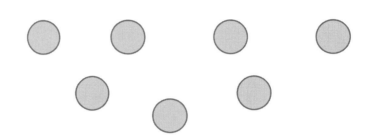

This probe is a hands-on variation that can be used with kindergarten students using a smaller quantity than the original probe that may also reveal what the student understands about initial place-value ideas.

Materials: 12 counters, dry-erase board or paper and marker

Instructions:

1. Place the 12 counters on the table in a random pattern in front of the student.
2. Ask the student to count the collection and tell you how many counters there are, and have them write the number if they are able, or you write the number they say for them.
3. Now, ask them to show you from the counters how many this digit means from the collection. *It is important for you not to say the digit name but to point to the digit in the ones place.*
4. Once the student has shown you that digit, continue to ask about the digit in the tens place without saying the digit name. *How much does this mean from the collection?*

Assessment:

☐ The goal is for the student to recognize that the total number of the collection is the same thing as the number name attached to the collection (*cardinality*).
☐ If the collection is too large still for a student, this probe can be used with a single-digit quantity to developmentally assess the concept of cardinality before the size of the numbers used by students become too large, too fast.

PROBE 2: WHAT'S THE NUMBER?

(Student Interview Task)

A)

```
10032
```

B)

```
132
```

C)

```
1032
```

Directions:

Show the calculators above to the student, and ask her to circle the calculator that shows the number you say, *"One hundred thirty-two."*

"How did you decide on calculator _____?"

Record reasoning:

TEACHER NOTES: WHAT'S THE NUMBER?

Questioning for Student Understanding

Can students translate numbers from verbal to symbolic representation?

K–2	3–5

Uncovering Understanding

Place Value: Whole Numbers Content Standard: Number and Operation

Note: This probe is an individual interview.

Variation/Adaptation: What's the Number? (using different numbers)

Examining Student Work

Students answers may reveal *misunderstandings* regarding the verbal name of a number and the visual representation of that number. A common misunderstanding may be revealed because students can identify a three-digit number that is written; however, when asked to create the number on a calculator or to write the number said, they have difficulty with the number of digits they use to represent the number.

- Choice A or C: Students who chose calculator A, are creating the number by initially including the 100 and then the 32. They are accounting for the word 100 as a whole and then including the 32. Calculator C similarly includes one less zero as students hear 100 and 32, so the zero becomes the "and."
- Choice B: Students who chose calculator B understand that a number in the hundreds is a three-digit number, and the zeros are replaced with the 32.

Seeking Links to Cognitive Research

During preschool and elementary-school years, children develop meanings for number words in which sequence, count, and cardinal meanings of number words become increasingly integrated. (AAAS, 1993, p. 350)

Concrete models can help students represent numbers and develop number sense; they can also help bring meaning to students' use of written symbols and can be useful in building place-value concepts. (NCTM, 2000, p. 80)

Understanding place value involves building connections between key ideas of place value—such as quantifying sets of objects by grouping by ten and treating the groups as units—and using the structure of the written notation to capture this information about groupings. Different forms of representation for quantities, such as physical materials and written symbols, highlight different aspects of the grouping structure. Building connections between these representations yields a more-coherent understanding of place value. (NCTM, 2003, p. 110)

The written place-value system is an efficient system that lets us write large numbers, but it is also abstract and misleading: The numbers in every position look the same. To understand the meaning of the numbers in the various positions, first- and second-grade children need experience with some kind of visual size-quantity supports, manipulatives, or drawings that show tens to be collections of 10 ones and show hundreds to be simultaneously 10 tens and 100 ones, and so on. (Clements & Sarama, 2004, p. 125)

Learning to communicate mathematically is one of the goals of the mathematics curriculum. One way to communicate mathematical ideas is through the use of symbols. (NCTM, 2003, p. 286)

In grades K–2: It is absolutely essential that students develop a solid understanding of the base-ten numeration system and place-value concepts by the end of Grade 2. (NCTM, 2000, p. 106)

Teaching Implications

To support a deeper understanding for students in the primary grades in regard to written and symbolic notation, the following are instructional ideas and questions to consider in conjunction with the research.

Focus Through Instruction

- Focus on uncovering students' thinking as they work with concrete materials by asking questions that elicit students' thinking and reasoning.
- Using base-10 language for numbers helps students to hear the place value and recognize the digit that refers to that place.
- Explicitly write numbers while saying the digit name as it relates to its position.
- Use of a calculator helps to develop and reinforce place-value concepts as students interact with the digits and words used along with the visual number formed.
- Use interactive technology to support student understanding. View an example activity from NCTM's Illuminations (2000–2010) site at http://illuminations.nctm.org/ActivityDetail.aspx?ID=73 (see information about using interactive applets in Chapter 1, page 22).

Questions to Consider (when working with students as they develop an understanding of place value)

- When composing and decomposing numbers, do students refer to the quantity of the digits based on their positions in the number?
- Do students recognize the relationship between a quantity and its number name?
- Do they refer to the digits in the tens and hundreds places as a single digit, or do they keep their value whole when they compose and decompose numbers?
- Can students generalize the pattern of tens to count and build numbers over 100?
- Are students able to confidently show how to "build a number" based on hearing the number aloud?

Teacher Sound Bite

I have to admit that when I initially looked at the probe What's the Number? I believed that my students would have no problem in completing it successfully. We had been working with numbers in the hundreds up to 999 for some time. We have used a 100s chart and built three-digit numbers with linking cubes while practicing saying the number names. The results as I sat with each student revealed that there was not the explicit connection between being able to say a number and reading and building a number by the number name. The results of my class provided me the insight to be more explicit in my instruction to provide multiple opportunities that involve reading, writing, using, building, and comparing numbers from two-digit to three-digit numbers.

Curriculum Topic Study and Uncovering Student Thinking

What's the Number?

Keeley, P., & Rose, C. (2006). *Mathematics curriculum topic study: Bridging the gap between standards and practice.* Thousand Oaks, CA: Corwin. (Place Value, p. 130; Numbers and Number Systems, p. 128).

Related Probes: How Many Stars? (page 42); What Is the Value of the Place? (page 54)

Rose, C., Minton, L., & Arline, C. (2007). *Uncovering student thinking in mathematics: 25 formative assessment probes.* Thousand Oaks, CA: Corwin. (Building Numbers, p. 38).

Additional References for Research and Teaching Implications

Burns, M. (2000). *About teaching mathematics: A K–8 resource.* Sausalito, CA: Math Solutions. (p. 173).

Clements, D., & Sarama, J. (2004). *Engaging young children in mathematics: Standards for early childhood mathematics education.* Mahwah, NJ: Lawrence Erlbaum. (p. 125).

National Council of Teachers of Mathematics. (2000). *Principles and standards for school mathematics.* Reston, VA: Author. (pp. 106, 152).

National Council of Teachers of Mathematics. (2002b). *Putting research into practice in the elementary grades.* Reston, VA: Author. (p. 286).

Student Responses to What's the Number?

Sample Responses for Choice A or C

Student 1: "I picked A because it is the right one. 10,032 is right."

Student 2: "I think C is right because you can make 132 with the one hundred, 0, and 32."

Sample Responses for Choice B

Student 3: "B is the right one because that is how you write 132. You don't need the zeroes to make 132."

Student 4: "132 is made of 100, 30, and 2, so it becomes 132, and that means that B is correct."

PROBE 2A: VARIATION: WHAT'S THE NUMBER?

(Student Interview Task)

1.

A)
```
10032
```

B)
```
132
```

C)
```
1032
```

Show the calculators above to the student, and ask him or her to circle the calculator that shows the number you say, *"One hundred thirty-two."* Ask, *"How did you decide on calculator _____?"*
Record reasoning:

2.

A)
```
1005
```

B)
```
105
```

C)
```
100.5
```

Show the calculators above to the student, and ask him or her to circle the calculator that shows the number you say, *"One hundred thirty-two."* Ask, *"How did you decide on calculator _____?"*
Record reasoning:

PROBE 3: WHAT IS THE VALUE OF THE PLACE?

<table>
<tr><td colspan="2" align="center"># 749</td></tr>
<tr><td colspan="2">Circle all the true statements about this number, and explain your thinking.</td></tr>
<tr><td>A) 7 tens and 49 ones</td><td>B) 749 ones</td></tr>
<tr><td>C) 7 hundreds 409 ones</td><td>D) 74 tens and 9 ones</td></tr>
<tr><td>E) 7 hundreds and 49 ones</td><td>F) 6 hundreds and 149 ones</td></tr>
</table>

TEACHER NOTES: WHAT IS THE VALUE OF THE PLACE?

Questioning for Student Understanding

Can students choose all correct values of various digits of a given number?

K–2	3–5

Uncovering Understanding

Place Value: Whole Numbers Content Standard: Number and Operation

Examining Student Work

Students answers may reveal *misconceptions* regarding the value of the "place" rather than the "face" value of the digit. A common misconception may be revealed as students refer to a digit by its face value rather than its place value and have difficulty decomposing multidigit numbers by their place value and associating compatible quantities.

- Choice A: Students who chose A, B, D, and E likely looked for the 7, 4, and 9 in the correct order. These students are looking at the individual digits and their order to determine the value of the number. They are not using an understanding of the value of each place in a three-digit number.
- Choice B: Students who chose B, D, E, and F correctly identified equivalent values of 749, demonstrating an understanding of the different ways quantity can be expressed based on our base-10 number system.

Seeking Links to Cognitive Research

Making a transition from viewing "ten" as simply the accumulation of 10 ones to seeing it both as 10 ones *and* as 1 ten is an important first step for students toward understanding the structure of the base-ten number. (NCTM, 2003, p. 33)

Concrete models can help students represent numbers and develop number sense; they can also help bring meaning to students' use of written symbols and can be useful in building place-value concepts. (NCTM, 2000, p. 80)

Children develop an understanding of the base-ten numeration system and place-value concepts (at least to 1,000). Their understanding of base-ten numeration includes ideas of counting in units and multiples of hundreds, tens, and ones, as well as a grasp of number relationships, which they demonstrate in a variety of ways, including comparing and ordering numbers. They understand multi digit numbers in terms of place value, recognizing that place-value notation is a shorthand for the sums of multiples of powers of 10 (e.g., 853 as 8 hundreds + 5 tens + 3 ones). (NCTM, 2006, p. 14)

Place value is foundational to all work with whole numbers and decimal numbers, and the lack of understanding of place value leads to most errors when computing with these numbers. (Sowder & Nickerson, 2010, p. 23)

Building on their work in Grade 2, students extend their understanding of place value to numbers up to 10,000 in various contexts. Students also apply this understanding to the task of representing numbers in different equivalent forms (e.g., expanded notation). (NCTM, 2006, p. 15)

Teaching Implications

To support a deeper understanding for students in the primary grades in regard to our number system, the following are instructional ideas and questions to consider in conjunction with the research.

Focus Through Instruction

- Notice what students do computationally that reveals what they understand about place value.
- Choose a variety of numbers and problem types to provide students with opportunity to use their knowledge of place value to develop a variety of strategies.
- Emphasize place value ideas by asking questions that support the base-10 number system (e.g., "What is 10 more? What is 10 less? What's 100 more? What is 100 less?").
- Use of calculators can develop and reinforce place-value concepts.
- Encourage the practice of decomposing numbers by their place-value quantities.

Questions to Consider (when working with students as they build and extend their understanding of place value)

- When composing and decomposing numbers, do students refer to the quantity of the digits based on their positions in the number?
- Do students recognize the relationship between a quantity and its number name?
- Do student strategies vary based on what they understand about the numbers involved in a problem, or do they apply the same strategy regardless of the numbers?
- Are students using number names appropriately when computing?
- Can students decompose numbers by place values other than just a single place? (E.g., 357 is how many tens? How many whole tens?)

Teacher Sound Bite

I was so interested to see what my students would do with this probe. As my teaching partner and I tried to predict our student responses, it was obvious that we both assumed they had a strong foundation to be able to choose the correct statements on the probe. To our surprise, 75% of our students had at least one incorrect response to the probe. From the responses, we noted that a consistent misunderstanding for our students was the idea that the number of tens in the number needed to include the hundreds in order for the answer to be correct. We had spent a lot of time breaking numbers apart by individual places but not in terms of the total number of tens in a three-digit number. We have changed our focus to include addressing all ways of decomposing and recomposing numbers by place value.

Additional References for Research and Teaching Implications

National Council of Teachers of Mathematics. (2000). *Principles and standards for school mathematics.* Reston, VA: Author. (p. 80–84).

National Council of Teachers of Mathematics. (2003). *Research companion to principles and standards for school mathematics.* Reston, VA: Author. (pp. 33–42).

National Council of Teachers of Mathematics. (2006). *Curriculum focal points for prekindergarten through grade 8 mathematics: A quest for coherence.* Reston, VA: Author. (pp. 14, 15).

Sowder J. L., & Nickerson, S. (2010). *Reconceptualizing mathematics for elementary school teachers.* Boston: W. H. Freeman. (p. 23).

Curriculum Topic Study and Uncovering Student Thinking

What Is the Value of the Place?

Keeley, P., & Rose, C. (2006). *Mathematics curriculum topic study: Bridging the gap between standards and practice.* Thousand Oaks, CA: Corwin. (Place Value, p. 130; Addition and Subtraction, p. 111).

Related Probes: How Many Stars? (page 42); What's the Number? (page 49)

Rose, C., Minton, L., & Arline, C. (2007). *Uncovering student thinking in mathematics: 25 formative assessment probes.* Thousand Oaks, CA: Corwin. (Building Numbers, p. 38).

Student Responses to What Is the Value of the Place?

Sample Responses: Choice A

Student 1: "I looked for the ones that had the 7, 4, and 9, so A, B, D, and E are the correct ones because the number is 749."

Student 2: "749 is made of 749 ones, so that is why I picked it."

Sample Responses: Choice B

Student 3: "I noticed that B was all of the ones for the number, and D was the 74 tens and ones, and E was the 7 hundreds and then 49, but F was tricky because you had to add the 600 + 149 to get 749."

Student 4: "I picked B, D, E, and F because each one when added together equals 749. A is 119, and C is 1,109, so they are not correct."

PROBE 4: WHAT NUMBER IS THAT?

1.

0   A   100

What number do you think the A represents? Explain your thinking.

_____

_____

_____

_____

_____

_____

2.

12   84

Put a mark where you think 30 would go on the number line. Explain your thinking.

_____

_____

_____

_____

_____

_____

TEACHER NOTES: WHAT NUMBER IS THAT?

<u>Q</u>uestioning for Student Understanding

Are students able to estimate a value on a number line given the value of the endpoints?

K–2	3–5

<u>U</u>ncovering Understanding

Place Value: Whole Numbers Content Standard: Number and Operation

Variation/Adaptation:

- What Number Is That? Smaller Number Range

<u>E</u>xamining Student Work

Students answers may reveal *misunderstandings* regarding the relative distance between whole numbers when shown on a number line. A common misunderstanding may be revealed as students choose to identify a location on a number line relating to just one of the endpoints without consideration of the other. Often, students count back by ones from the end point to land on the specific location using a finger or pencil width as the unit.

- Correct answers are, for 1, between 70 and 75 and, for 2, 30 is close to the midpoint (36 is the midpoint between 12 and 84). Students who answered both correctly have an understanding of the position of numbers in the sequence and of their relative relationship within different starting and ending points.
- Incorrect responses: Students who said 90 to 94 typically counted back from 100 and laid their finger or pencil down and counted by ones until they landed on the A. Students who placed the 30 mark closer to the 84 counted up from the 12 in a similar fashion and got fairly close to the 84. They do not have a sense of the relative distance between sets of numbers and therefore cannot apply the concept accurately.

Seeking Links to Cognitive Research

During preschool and elementary-school years, children develop meanings for number words in which sequence, count, and cardinal meanings of number words become increasingly integrated. (AAAS, 1993, p. 350)

Using the number line in addition to other representations can strengthen student understanding. (Clarke, Roche, & Mitchell, 2008, p. 374)

Students should understand that written representations of mathematical ideas are an essential part of learning and doing mathematics. It is important to encourage students to represent their ideas in ways that make sense to them even if their first representations are not conventional ones. It is also important that they learn conventional forms of representation to facilitate both their learning of mathematics and their communication with others about mathematical ideas. (NCTM, 2000, p. 67)

Concrete models can help students represent numbers and develop number sense; they can also help bring meaning to students' use of written symbols and can be useful in building place-value concepts. (NCTM, 2000, p. 80)

A number line has several advantages. First, it shows the distance from zero (or the absolute value of the number). In addition, it is an excellent tool for modeling the operations. Jumps can be shown in the same way as with the whole numbers and fractions. Students can see that integer moves to the left go to smaller numbers, and moves to the right go to larger numbers (Van de Walle, 2010, p. 482).

Keep in mind that the labeling of the points for numbers on a number line comes about because of their distances from 0, a starting point. But + 2, could be thought of as any jump of length 2 units toward the right but starting anywhere (not necessarily at the 0 point), just as − 3 could be thought of as any jump of length 3 units, but toward the left and starting anywhere. Thinking of a signed number as a description of a jump size rather than only as a point on the number line has value in working with addition and subtraction of signed numbers. (Sowder & Nickerson, 2010, p. 181)

Teaching Implications

To support a deeper understanding of the patterns embedded in our number system and students' ability to build a mental model, consider the following ideas from research.

Focus Through Instruction

- Students should be familiar with a linear representation of a number on a number line before they are introduced to an empty number line.
- Students should be able to count by tens both on and off the decade and jump across tens.
- Students should record only the relevant numbers on the number line as they represent their thinking.
- Consider that the number is a geometric model of all real numbers.
- Unlike counters, which model only counting, the number line models measurement; so, beginning at zero is an important idea for both linear measurement and numbers.
- Use interactive technology to support student understanding.
- View example activities, Number Line Arithmetic and Place Value Number Line from National Library of Virtual Manipulatives site at http://nlvm.usu.edu/en/nav/topic_t_1.html (see information about using interactive applets in Chapter 1, page 22).

Questions to Consider (when working with students as they develop an understanding of the number line as a model to represent their thinking)

- Do students have a clear sense of the magnitude of numbers as they relate to each other out of the rote sequence?
- Do students recognize the relationship between a quantity and its location?
- Do students understand that when counting the first object is "one," but when we measure using a ruler (number line), we line one end up against the zero mark on the ruler?
- Do students notice both the sequence of the numbers on a number line and the intervals used so that they can use the number line flexibly?

Teacher Sound Bite

I noticed from using this probe that while my students were able to place given numbers in the relative position with given endpoints, they were not as successful when they were asked to provide the numbers for the given points. My students also consistently used only one of the endpoints to base their decisions on a number to assign and the relationship between endpoints was never shared as a strategy by any of my students.

Curriculum Topic Study and Uncovering Student Thinking

What Number Is That?

Keeley, P., & Rose, C. (2006). *Mathematics curriculum topic study: Bridging the gap between standards and practice.* Thousand Oaks, CA: Corwin. (Numbers and Number Systems, p. 128; Comparing and Ordering Numbers, p. 114).

Rose, C., Minton, L., & Arline, C. (2007). *Uncovering student thinking in mathematics: 25 formative assessment probes.* Thousand Oaks, CA: Corwin.

Additional References for Research and Teaching Implications

National Council of Teachers of Mathematics. (2000). *Principles and standards for school mathematics.* Reston, VA: Author. (pp. 67, 80).

Sowder, J. L., & Nickerson, S. (2010). *Reconceptualizing mathematics for elementary school teachers.* Boston: W. H. Freeman. (p. 181).

Van de Walle, J. A. (2010). *Elementary and middle school mathematics* (7th ed.). Boston: Pearson. (p. 482).

Student Responses to What Number Is That?

Sample Responses: Incorrect

Student 1: "I say it is 94 because I put my finger at 100 and counted back to the letter A, and that was 94."

"I counted from 12 to 30 and put the X." (The mark is about an inch from the 84 endpoint.)

Student 2: "94, because it will take 6 to get from A to 100."

"I put the X on the spot where I landed. 30 is pretty close to 12, so it goes close to 12 if you count by ones."

Student 3: "I think the answer is 70 because halfway is 50 and 2 jumps of 10 from there would be 70. The A is closer to the middle than the 100, so 70 makes sense."

"30 would be about halfway between 12 and 84 because I made the 12 a 10 and the 84 an 80, which means there are 70 numbers and half of 70 is 35, so about halfway is an estimate."

Student 4: "Halfway between 0 and 100 is 50, and the A is a little closer to the 50 than the 100, so I say 70 is correct."

"You have to see that there is 72 between 12 and 84, so 36 is halfway, and 30 is a little bit closer to the 12, so the X would be a little less than halfway closer to the 12 end."

PROBE 4A: VARIATION: WHAT NUMBER IS THAT?

4a

$$\longleftrightarrow \begin{array}{ccccccc} -3 & -2 & -1 & 0 & A & 2 & 3 \end{array} \longrightarrow$$

1.

$$0 \qquad\qquad\qquad\qquad\qquad A \qquad\quad 10$$

What number do you think the *A* represents? Explain your thinking.

_____

_____

_____

_____

_____

_____

2.

$$8 \qquad\qquad\qquad\qquad\qquad\qquad\qquad\qquad 58$$

Put a mark where you think 30 would go on the number line? Explain your thinking.

_____

_____

_____

_____

_____

PROBE 5: HUNDRED CHART CHUNKS

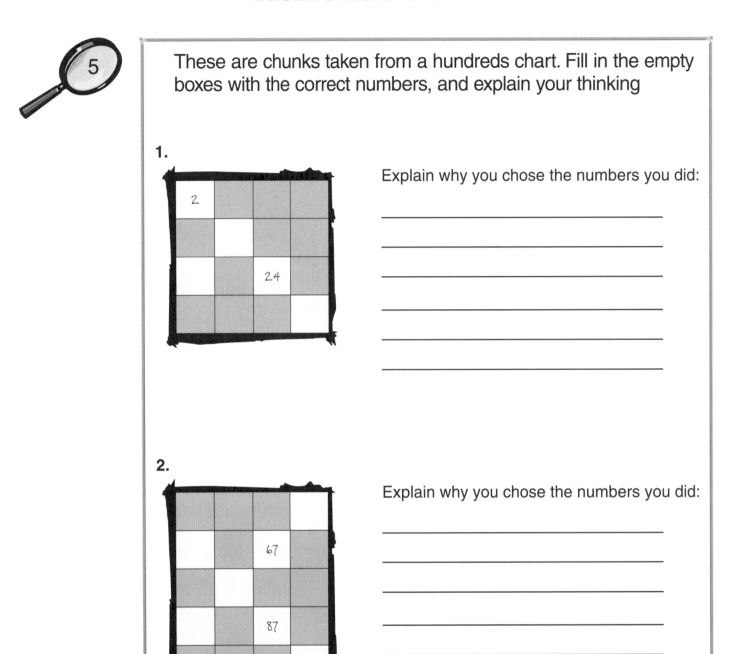

These are chunks taken from a hundreds chart. Fill in the empty boxes with the correct numbers, and explain your thinking

1.

Explain why you chose the numbers you did:

_____

_____

_____

_____

_____

_____

2.

Explain why you chose the numbers you did:

_____

_____

_____

_____

_____

_____

TEACHER NOTES: HUNDRED CHART CHUNKS

<u>Q</u>uestioning for Student Understanding

Given a portion of the 100s chart, can students correctly fill in the missing values?

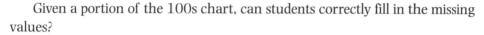

K–2	3–5

<u>U</u>ncovering Understanding

Place Value: Whole Numbers Content Standard: Number and Operation

Variation/Adaptation:

- Hundreds Chart Chunks: One Example

<u>E</u>xamining Student Work

Students answers may reveal *misunderstandings* regarding the intended structure of the 100s chart and a student's ability to use the patterns of the number system to support math reasoning. Student dependence on a tool should be monitored to ensure that the tool is meaningful, generalizable, and consistent with the concepts being addressed.

- The correct answers for Question 1 are 13, 22, and 35; the correct answers for Question 2 are 58, 65, 76, and 85. These responses demonstrate an understanding of the patterns of the chart and support a student's knowledge of the structure of our number system without the student being dependent on a completed chart.
- Students who completed the chunks with patterns of their own choosing (took jumps of 14 from the known number to missing number, completed by odd or even numbers, etc.) tried to make sense of creating a pattern, but they failed to either recognize, transfer, or use the structure of the 100s-chart model.

<u>S</u>eeking Links to Cognitive Research

Research indicated that students' experiences using physical models to represent hundreds, tens, and ones can be effective if the materials help them think about how to combine quantities and, eventually, how these processes connect with written procedures. (National Research Council [NRC], 2001, p. 96–99)

Concrete models can help students represent numbers and develop number sense; they can also help bring meaning to students' use of written symbols and can be useful in building place-value concepts. (NCTM, 2000, p. 80)

Teachers should try to uncover students' thinking as they work with concrete materials by asking questions that elicit students' thinking and reasoning. In this way, teachers can watch for students' misconceptions, such as interpreting the 2 tens and 3 ones merely as five objects. (NCTM, 2000, p. 80)

Concrete materials can be an effective aid to students' thinking and to successful teaching. But, effectiveness is contingent on what one is trying to achieve. To draw maximum benefit from students' use of concrete materials, the teacher must continually situate her or his actions with the question "What do I want my students to understand?" (NCTM, 2002b, p. 249)

Understanding place value involves building connections between key ideas of place value—such as quantifying sets of objects by grouping by ten and treating the groups as units—and using the structure of the written notation to capture this information about groupings. Different forms of representation for quantities, such as physical materials and written symbols, highlight different aspects of the grouping structure. Building connections between these representations yields a more-coherent understanding of place value. (NCTM, 2003, p. 110)

In Grades K–2: It is absolutely essential that students develop a solid understanding of the base-ten numeration system and place-value concepts by the end of Grade 2. (NCTM, 2000, p. 106)

In Grades 3–5: Students should be computing fluently with whole numbers. Computational fluency refers to having efficient and accurate methods for computing. Students exhibit computational fluency when they demonstrate flexibility in the computational methods they choose, understand and can explain these methods, and produce accurate answers efficiently. The methods that a student uses should be based on mathematical ideas including the structure of the base-ten number system. (NCTM, 2000, p. 152)

Teaching Implications

To support a deeper understanding for students in elementary school in regard to place value, the following are ideas to consider when working with students.

Focus Through Instruction

- Physical models for base-10 concepts can play a key role in helping children develop the idea of numbers having both a single identity and a unit identity.
- A 100s chart, no matter the style, is a pedagogical device used to support visualization of the structure of our number system to practice both the naming and writing of whole numbers.
- Use of proportional base-10 models encourages students to think about how they are using numbers.
- Base-10 number knowledge results from an ability to count, to make groupings, and to understand place value at a deeper level than simply moving on a chart.
- Students can foster understanding of place value through their experiences with combining and comparing numbers using a variety of models to confirm multiple ways of representing the same quantity or operation.
- Use interactive technology to support student understanding.
- View an example activity from National Library of Virtual Manipulatives site at http://nlvm.usu.edu/en/nav/topic_t_1.html (see information about using interactive applets in Chapter 1, page 22).

Questions to Consider (when working with students as they extend an understanding of place value and the patterns of our base-10 number system)

- Do students understand the equivalence of one group and the discrete units that constitute it?
- How do students use the 100s chart to solve problems?
- Do students use this chart in a way such that it connects with their existing knowledge and is therefore meaningful to them?
- Can students employ strategies of using our number system both with and without the use of a 100s chart equally as effectively?
- Can students generalize the pattern of tens to count and build numbers over 100?

Teacher Sound Bite

My students have been using the 100s chart since Grade 1. They have had many opportunities to use it for finding patterns, playing games, computing, and more. They have copies in their hardcover reference books, we have a large 200 chart on our wall, and our math program refers to it often. I wanted to know if my students could "let go" of the chart as I questioned their conceptual understanding of the chart based on how often students' appeared to rely on the chart to solve computation problems. I was amazed by the difficulty my students had in completing the "chunks" from a 100s chart. Without the whole chart visible, students completed the blank spaces with either consecutive numbers, or they made up a pattern that they thought completed the chunk. What was more interesting to me was that, although my students have been circling, coloring, and reciting the patterns of the horizontal and vertical number grid since grade one, they did not appear to understand how the chart is set up. This was a huge eye opener for me, making me reconsider the models used in my program to ensure that my use of the model is consistent with intent of the mathematics that underlie the model. Given this information, I have transformed my instructional use of the 100s chart as I am more focused on moving students away from the chart in ways that help them to build a mental model of the base-10 system that works for them.

Additional References for Research and Teaching Implications

Burns, M. (2000). *About teaching mathematics: A K–8 resource.* Sausalito, CA: Math Solutions Publications. (p. 173).

National Council of Teachers of Mathematics. (2000). *Principles and standards for school mathematics.* Reston, VA: Author. (pp. 80, 106, 152).

National Council of Teachers of Mathematics. (2002). *Putting research into practice in the elementary grades.* Reston, VA: Author. (p. 249).

National Council of Teachers of Mathematics. (2003). *Research companion to principles and standards for school mathematics.* Reston, VA: Author. (p. 110).

National Research Council. (2001). *Adding it up: Helping children learn mathematics.* Washington, DC: National Academy Press. (pp. 96–99).

Curriculum Topic Study and Uncovering Student Thinking

Hundred Chart Chunks

Keeley, P., & Rose, C. (2006). *Mathematics curriculum topic study: Bridging the gap between standards and practice.* Thousand Oaks, CA: Corwin. (Place Value, p. 130; Numbers and Number Systems, p. 128).

Related Probe: What Is the Value of the Place? (page 54)

Rose, C., Minton, L., & Arline, C. (2007). *Uncovering student thinking in mathematics: 25 formative assessment probes.* Thousand Oaks, CA: Corwin. (Building Numbers, p. 38).

Student Responses to Hundred Chart Chunks

Sample Responses

Student 1: "19 because 24 − 5 = 19, and 24 − 2 = 22."

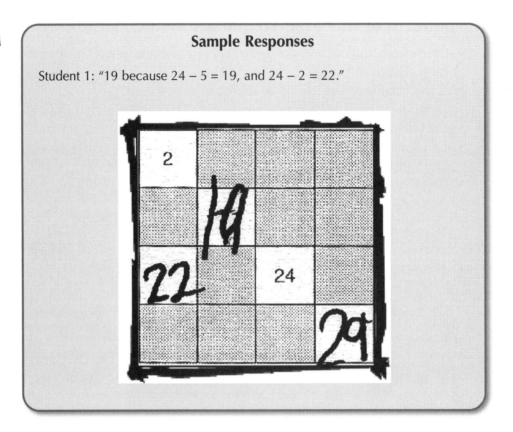

Sample Responses

Student 2: "64 because I counted backwards, 66, 65, 64."

Sample Responses

Student 3: "If you start at 2 and add 10, you would be at 12; add one, you'd be at 13; and add 10, you'd be at 23; minus 1, you'll have 22; add two, 24; add 1, 25; add 10, is 35."

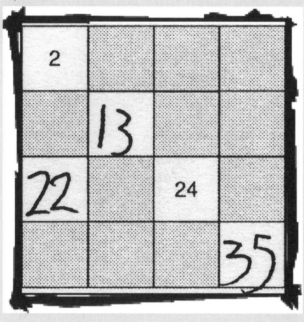

Sample Responses

Student 4: "If you start at 67 minus 10, you'll have 57; add 1, you're at 58; add 20, you're at 78; minus 2, 76; add 10, 86; minus 1, 85; minus 10, 95; add 3, 98."

PROBE 5A: VARIATION: HUNDRED CHART CHUNKS

These are chunks taken from a hundred chart.
Fill in the empty boxes with the correct numbers.

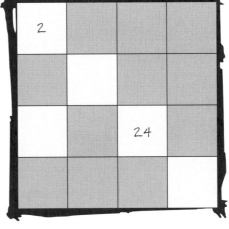

Explain why you chose the numbers you did:

_____

_____

_____

_____

_____

_____

_____

PROBE 6: WHAT IS THE VALUE OF THE DIGIT?

Circle all of the statements that are true for the number 2.13.	
Statement	**Explanation (why circled or not circled)**
A) There is a 3 in the ones place.	
B) There is a 2 in the ones place.	
C) There are 21.3 tenths.	
D) There are 13 tenths.	
E) There is a 1 in the tenths place.	
F) There is a 3 in the tenths place.	
G) There are 21 hundredths.	
H) There are 213 hundredths.	

TEACHER NOTES: WHAT IS THE VALUE OF THE DIGIT?

Questioning for Student Understanding

Can students choose all correct values of various digits of a given decimal?

K–2	3–5

Uncovering Understanding

What Is the Value of the Digit? Content Standard: Number and Operation

Variation/Adaptation:
- What Is the Value of the Digit? Card Sort

Examining Student Work

The distracters may reveal lack of *procedural* and *conceptual understanding* of the place-value positions in the base-10 number system as well as *conceptual understanding* of the value of individual digits in relationship to the represented number.

- The correct answers are B, C, E, and H: Students who choose B and E understand they can name the value of the place of digits within a number. By choosing C and H, these students also demonstrate a conceptual understanding of the relationship between the value of the places and the number represented by the specific combination of digits. (See Student Responses 1.) Many students can accurately identify B and E but exclude choices C and H. These students demonstrate a procedural understanding of naming digits in specific places but lack conceptual understanding of the value of the digits in relationship to the represented number. (See Student Response 2.)
- Distracters A and F: Students who choose A or F are demonstrating a *common error* related to the place value of the digits in a decimal. These students *overgeneralize* from their work with whole numbers by thinking the ones place is always the right-most number, or they think that because the tens place is second from the left of the decimal point, the tenths place must be second from the right of the decimal point. (See Student Response 3.)
- Distracters D and G: Students who choose D and/or G demonstrate the *common error* of considering only the numbers to the left of the identified place. This typically is related to a lack of conceptual understanding of the relationship of each digit to the size of the number. (See Student Response 4.)

Seeking Links to Cognitive Research

Elementary and middle school students may have limited ability with place value. Sowder (1992) reports that middle school students are able to identify the place values of the digits that appear in a number, but they cannot use the knowledge confidently in context. (AAAS, 1993, p. 350)

Upper elementary students often do not understand that decimal fractions represent concrete objects that can be measured by units, tenths of units, hundredths of units, and so on (Hiebert, 1992). (AAAS, 1993, p. 350)

Students have little understanding of the value represented by each of the digits of a decimal number or know the value of the number is the sum of the value of its digits. (AAAS, 1993, p. 350)

Student errors suggest students interpret and treat multi-digit numbers as single-digit numbers placed adjacent to each other, rather than using place-value meanings for the digits in different positions. (AAAS, 1993, p. 358)

Students in grades 3–5 should use models and other strategies to represent and study decimal numbers. (NCTM, 2000, p. 150)

Research has confirmed that a solid conceptual grounding in decimal numbers is difficult for students to achieve. The similarities between the symbol systems for decimals and whole numbers lead to a number of misconceptions and error types. Grasping the proportional nature of decimals is particularly challenging. (NRC, 2005, p. 332)

In their struggle to find meaning with decimal place value, students display a variety of difficulties, including language difficulties. They may say "tens" for "tenths" and "hundreds" for "hundredths" (Resnick et al., 1989). (NCTM, 1993a p. 140)

Teaching Implications

To support a deeper understanding for students in secondary grades in regards to number and operations, the following are ideas and questions to consider in conjunction with the research.

Focus Through Instruction

- The foundation of students' work with decimal numbers must be an understanding of whole numbers and place value.
- Representing numbers with various physical materials helps build understanding of place value.
- Decimal instruction should build on students' encounters with fractions, decimals, and percentages in prior grades, and it should be situated in familiar contexts.

- Experiences with a variety of models, such as fraction strips, number lines, 10×10 grids, area models, and objects offer concrete representations of abstract ideas.
- Proper use of the terminology can help students connect language and place-value concepts (e.g., 2 and 3 tenths rather than 2.3).

Questions to Consider (when working with students as they grapple with the idea of place value)

- Do students understand the role of the decimal point and the relationship among the digits in the ones, tenths, and hundredths places?
- Are students able to move beyond just naming digits in various places to the value of the digit in relationship to the number?
- Are students able to show various concrete representations of the number (e.g., using base-10 blocks or other visual representations)?

Teacher Sound Bite

I have to admit that this probe made me stop and think about what the correct answers would be when we reviewed the items at our Grade 4 teacher meeting. I used the probe before introducing an activity to provide a context to the concept. We used our coins-and-dollars manipulative to talk about various ways to get a total dollar amount. After our hands-on activity, I gave them the probe again and saw an improvement with the majority of the students. Some, though, still needed to use the manipulative before being able to complete the probe.

Curriculum Topic Study and Uncovering Student Thinking

What Is the Value of the Digit?

Keeley, P., & Rose, C. (2006). *Mathematics curriculum topic study: Bridging the gap between standards and practice.* Thousand Oaks, CA: Corwin. (Place Value, p. 130).

Rose, C., Minton, L., & Arline, C. (2007). *Uncovering student thinking in mathematics: 25 formative assessment probes.* Thousand Oaks, CA: Corwin. (Comparing Decimals, p. 55).

Additional References for Research and Teaching Implications

American Association for the Advancement of Science. (1993). *Benchmarks for science literacy.* New York: Oxford University Press. (pp. 350, 358–359).

National Council of Teachers of Mathematics. (2002b). *Putting research into practice in the elementary grades.* Reston, VA: Author. (pp. 113–118).

National Research Council. (2005). *How students learn: Mathematics in the classroom.* Washington, DC: National Academy Press. (pp. 309–343).

Van de Walle, J. A. (2007). *Elementary and middle school mathematics* (6th ed.). Boston: Pearson. (pp. 335–337).

Student Responses to What Is the Value of the Digit?

Sample Responses: B, C, E, and H

Student 1: "B and E: I circled the place and looked at the number to see if it matched. C and H—C: There are 1 tenths (.1) and 20 tenths (2) and 3 hundredths which is 3/10 of a 1 tenth. H: There are 200 hundredths (2), 1 hundredth (.1) and 3 hundredths (.03), so that makes 213 hundredths."

Sample Responses: B and E

Student 2: "B because I know the ones place is 1 spot before decimal. And E because I know the tenths spot is one place after the decimal."

Sample Responses: Inclusion of A and/or F

Student 3: "F. Tenths is like tens."

Sample Responses: Inclusion of D and/or G

Student 4: "D because 3 is in tenths place, so .13 would mean 13 tenths. Don't worry about 2 because it is the whole-number part."

PROBE 6A: VARIATION: WHAT IS THE VALUE OF THE DIGIT? CARD SORT

A) There is a 3 in the ones place.	**E)** There is a 1 in the tenths place.
B) There is a 2 in the ones place.	**F)** There is a 3 in the tenths place.
C) There are 21.3 tenths.	**G)** There are 21 hundredths.
D) There are 13 tenths.	**H)** There are 213 hundredths.

4

Structure of Number Probes

Parts and Wholes and Equality

Figure 4.1   Chapter 4 Probes

Grade-Span Bar Key								
	Target for Instruction Depending on Local Standards							
	Prerequisite Concept and Field Testing Indicate Students May Have Difficulty							

Question	Probe	Grade Span						
Structure of Number: Parts and Wholes		K	1	2	3	4	5	
Do students understand the meaning of the equal sign?	Equal to 4? (page 78)				2–4		5	
Given a set model, are students able to define fractional parts of the whole?	Crayon Count (page 85)				2–4		5	
Given a whole, are students able to indentify when $\frac{1}{4}$ of the whole is shaded?	Is $\frac{1}{4}$ of the Whole Shaded? (page 91)					3–4	5	
Given an area model, are students able to define fractional parts of the whole?	Granola Bar (page 99)						4–5	
Are students able to choose equivalent forms of a fraction?	Is It Equivalent? (page 104)						4–5	
Do students use the "cancelling of zeros" shortcut appropriately?	Is It Simplified? (page 110)						4–5	

PROBE 7: EQUAL TO 4?

Circle only the math sentences where $\square = 4$.

A) $2 + 2 = \square - 3$

B) $9 - \square = 5$

C) $10 - 6 = \square$

D) $\square = 1 + 3$

E) $6 + 3 = \square + 5$

F) $3 + 1 = \square + 2$

Explain your choices:

TEACHER NOTES: EQUAL TO 4?

Questioning for Student Understanding

Do students understand the meaning of the equal sign?
Grade Level for Equal to 4?

K–2	3–5

Uncovering Understanding

Equal to 4? Content Standard: Algebra

Variation/Adaptation:

- Variation: Equal to 14? (Probe 7a) or Variation: Equal to 14? (Probe 7b)

Examining Student Work

The distracters may reveal *common errors* regarding equivalence as it relates to solving equations with an unknown variable. The concept of equality is a crucial idea for developing algebraic reasoning in young children.

- The correct answers are B, C, D, and E. Students who correctly choose these equations demonstrate a conceptual understanding of the idea of equivalence (i.e., balance). (See Student Response 1.)
- Distracter A and F and E. Students who *include* A and F and *exclude* E typically see the equals sign as representing the end of an equation or as a signal to give the answer to the expression before the equals sign. (See Student Response 2.)
- Distracter D. Students whose answers *include* B and C but *exclude* D do not have fully developed understanding of equality and often see the equation as being wrong because it is not written in a familiar format with the numerical expression to the left of the equal sign or only one missing value to the left of the equal sign. (See Student Response 3.)

Seeking Links to Cognitive Research

Children in the elementary grades generally think that the equals sign means that they should carry out the calculation that precedes it and that the number after the equals sign is the answer to the calculation. Elementary school children generally do not see the equals sign as a symbol that expresses the relationship "is the same as." (National Council of Teachers of Mathematics [NCTM], 2002b, p. 203)

Children must understand that equality is a relationship that expresses the idea that two mathematical expressions hold the same value. It is important for children to understand this idea for two reasons. First, children need this understanding to think about relationships expressed by number sentences. A second reason that understanding equality as a relationship is important is that a lack of such understanding is one of the major stumbling blocks for students when they move from arithmetic to algebra. (NCTM, 2002b, p. 206)

The notion of equality also should be developed throughout the curriculum. As a consequence of the instruction they have received, young students typically perceive the equals sign operationally, that is, as a signal to "do something." Instead, they should come to view the equals sign as a symbol of equivalence and balance. (NCTM, 2000, p. 39)

In grades K–2, equality is an important algebraic concept that students must encounter and begin to understand. A common explanation for the equals sign given by students is that "the answer is coming" but they need to recognize that the equals sign indicates a relationship—that the quantities on each side are equivalent. (NCTM, 2000, p. 94)

In grades 3–5, the idea and usefulness of a variable (represented by a box, letter, or symbol) should also be emerging and developing more fully. As students explore patterns and note relationships, they should be encouraged to represent their thinking. (NCTM, 2000, p. 161)

<u>T</u>eaching Implications

In order to support a deeper understanding for students in regard to equivalence, the following are ideas and questions to consider in conjunction with the research.

Focus Through Instruction

- Make direct comparisons between quantities by asking students about the relationship.
- Offer varied representations denoting equivalence using the equals sign.
- Use explicit language during instruction to refer to the equals sign as a relationship between the two numbers and/or expressions on opposite sides of the symbol.

- Use visual models that support the idea of equivalence (e.g., balance/seesaw).
- Embed symbolic representation and manipulation in instructional experiences to support sense making for students.
- Provide opportunities for students to make connections from symbolic notation to the representation of an equation.
- Use a balance and cubes to demonstrate equalities.
- Use interactive technology to have students model equivalence. View an example applet from NCTM's Illuminations (2000–2010) at http://illuminations.nctm.org/ActivityDetail.aspx?ID=10 (see information about using interactive applets in Chapter 1, page 22).

Questions to Consider (when working with students as they grapple with the idea of equivalence)

- Are students thinking about the equals sign only as "the answer is coming"?
- Can students use concrete objects to demonstrate equality?
- Can students articulate equality as two mathematics expressions that have the same value?
- Do students understand the unknown quantity can be represented in various positions (i.e., not always directly before or after the equals sign)?
- Are students representing the idea of equivalence through the development of their addition and subtraction strategies?

Teacher Sound Bite

I had students individually complete this probe without any prior instruction or discussion. We were to begin our supplemental algebra unit soon, so I expected there to be a good many misconceptions, which of course there were. In the past, students have done very well with the hands-on components of the unit but still had difficulty transferring their understanding to equation form. This year, I made sure to have students work on several related sets of problems; and after, we discussed how the written equations were similar or different based on the concrete modeling of the problem. I also ended each math session with an exit ticket that included two questions similar to that on the probe. At the end of the unit, I gave students the Equal to 4? probe and asked if they still agreed with their first choices and explanations. Students worked in groups to create a visual to show the answers and explanations. As a final activity, I gave Equal to 14? as a final individual assessment of progress.

<table>
<tr><td>

Curriculum Topic Study and Uncovering Student Thinking

Equal to 4?

Keeley, P., & Rose, C. (2006). *Mathematics curriculum topic study: Bridging the gap between standards and practice.* Thousand Oaks, CA: Corwin. (p. 194).

Related Elementary Probes:

Rose, C., Minton, L., & Arline, C. (2007). *Uncovering student thinking in mathematics: 25 formative assessment probes.* Thousand Oaks, CA: Corwin. (It's all About Balance, p. 90; Seesaw, p. 95).

</td></tr>
</table>

Additional References for Research and Teaching Implications

National Council of Teachers of Mathematics. (1993). *Research ideas for the classroom: Early childhood mathematics.* New York: Macmillan. (pp. 11–15).

National Council of Teachers of Mathematics. (2000). *Principles and standards for school mathematics.* Reston, VA: Author. (pp. 39, 94, 161).

National Council of Teachers of Mathematics. (2002). *Putting research into practice in the elementary grades: Readings from journals of the NCTM.* Reston, VA: Author. pp. 202–206.

National Research Council. (2001). *Adding it up: Helping children learn mathematics.* Washington, DC: National Academy Press. (pp. 261–263).

Student Responses to Equal to 4?

Sample Responses: B, C, D, and E

Student 1: "I made a T and looked to see if putting in 4 in the box and doing any math made both sides of the T the same number."

Sample Responses: Include A and F

Student 2: "2 + 2 = 4 − 6 on number line means go to left of 0 and land −2. 3 + 1 = 4 + 2 = 6." (Student did not explain noncircled sentences but did not circle E.)

Sample Responses: Include B and C, but Exclude D

Student 3: "I think D is 4, but it is a trick so I didn't add it as being right."

PROBE 7A: VARIATION: EQUAL TO 14?

Circle only the math sentences where □ = 14.

7a

A) $10 + 4 = \square - 3$	D) $\square = 14 + 5 - 6$
B) $18 - \square = 4$	E) $8 + 9 = \square + 3$
C) $\square - 6 = 13 - 5$	F) $20 - 5 = 11 + \square - 2$

Explain your choices:

PROBE 7B: VARIATION: EQUAL TO 14?

Circle only the math sentences where $a = 14$.

A) $10 + 4 = a - 3$	D) $a = 14 + 5 - 6$
B) $18 - a = 4$	E) $8 + 9 = a + 3$
C) $a - 6 = 13 - 5$	F) $20 - 5 = 11 + a - 10$

Explain your choices:

PROBE 8: CRAYON COUNT

There are 12 crayons in Sophie's box. There are blue, orange, and purple crayons.

- One-fourth ($\frac{1}{4}$) of the crayons are orange.

- One-fourth ($\frac{1}{4}$) of the crayons are blue.

- Half ($\frac{1}{2}$) of the crayons are purple.

Are there *more* purple or blue crayons in Sophie's box? _____

Are there *more* blue or orange crayons in Sophie's box? _____

How many of *each* color are in Sophie's box? _____

TEACHER NOTES: CRAYON COUNT

<u>Q</u>uestioning for Student Understanding

Given a set model, are students able to define fractional parts of the whole?

K–2	3–5

<u>U</u>ncovering Understanding

Fractional Parts Content Standard: Number and Operation

<u>E</u>xamining Student Work

Students' answers may reveal *overgeneralizations* regarding application of whole-number properties to working with fractions.

- The correct responses are purple; same; and 3 orange, 3 blue, and 6 purple. Students who answer as above are demonstrating their understanding of the relationships between the 3 colored crayons and the total of 12. They are using their knowledge of part-whole relationships.
- Students who answered with other quantities: Students who answer differently likely considered the total number of crayons without relating to the amount of the different colors. They most likely see all of the crayons as the total, much the same as in whole numbers, without considering the varying amounts indicated by the orange, blue, and purple colors. For example, some students focus on the fractions and then make the numbers fit (i.e., blue and orange are the same, so 2 + 2 = 4, so there must be 8 purple). They focus on solving the number sentence rather than applying all of the information from the original problem.

<u>S</u>eeking Links to Cognitive Research

One of the primary characteristics of students' informal knowledge of fractions is that students' informal solutions involve separating units into parts and dealing with each part as though it represents a whole number, as opposed to dealing with each part as a fraction. (Mack, 1990, p. 21)

Fractions make it possible to represent numbers between whole numbers. Fractions express numbers as an indicated division of two whole numbers. A fraction bar indicates the division. Some of the main ways that fractions are used in elementary school are as part of the whole, as quotient representations of rations, as measures, as individual numbers on a number line, and in computation. (Bay Area Mathematics Task Force, 1999, p. 59)

Fractions are a critical foundation for students, as they are used in measurement across various professions, and they are essential to the study of algebra and more advanced mathematics. This understanding must go well beyond recognizing that $\frac{3}{5}$ of a region is shaded. (Van de Walle, 2010, p. 287)

Understanding fractions means understanding all the possible concepts that fractions can represent. One of the commonly used meanings of fraction is part-whole, including examples when part of a whole is shaded. In fact, part-whole is so ingrained in elementary textbooks as the way to represent fractions, it may be difficult for you to think about what else fractions might represent. (Van de Walle, 2010, p. 287)

Multiplication and division situations in grades K–4 are often limited just to whole number situations. This limitation does not reflect reality. It also leads children to conclude falsely that multiplication always makes the answer larger and division always makes the answer smaller. This, in turn, reinforces some of the immature strategies used by children. Children need hands-on experiences such as sharing three candy bars among six children or sharing two-and-a-half pizzas among nine children. The experiences emphasize the need to represent what is actually happening in a situation. They also give children the necessary background for visualizing and connecting symbolic sentences that appear in later grades. (NCTM, 1993a, p. 105)

In grades K–2, in addition to work with whole numbers, young students should also have some experience with simple fractions through connections to everyday situations and meaningful problems, starting with the common fractions expressed in the language they bring to the classroom, such as "half." At this level it is more important for student to recognize when things are divided into equal parts than to focus on fraction notation. (NCTM, 2000, p. 82)

In grades 3–5, students should build their understanding of fractions as parts of a whole and as division. They will need to see and explore a variety of models of fractions, focusing primarily on familiar fractions such as halves, thirds, fourths, fifths, sixths eighths, and tenths. By using an area model in which part of a region is shaded, students can see how fractions are related to a unit whole, compare fractional parts of a whole, and find equivalent fractions. (NCTM, 2000. p. 150)

<u>T</u>eaching Implications

In order to support an initial understanding for elementary students regarding fractions, specifically with using part-to-whole relationships to find the missing whole, the following are ideas and questions to consider in conjunction with the research.

Focus Through Instruction

- For students to really understand fractions, they must experience fractions across many constructs.
- Three categories of models support ideas of fractions: area, length, and a set quantity. This probe utilizes the idea of a "set."
- Use of an area model allows students to see the part-to-whole relationship.
- Part-to-whole is one meaning of fractions and goes beyond shading a region.
- Support students to see how fractions and whole numbers are similar and different.
- Provide many opportunities to introduce students to fractions and situations that utilize fractions in a context that is developmentally appropriate.
- Use interactive technology to have students sort shapes by various attributes
- View an example applet from NCTM's Illumination's (2000–2010) at NCTM's Calculation Nation at http://calculationnation.nctm.org/Games/ (see information about using interactive applets in Chapter 1, page 22).

Questions to Consider (when working with students as they develop an initial understanding of fractions)

- Do students use what they know about whole numbers when working with fractions?
- Do students understand the difference between the numerator and denominator?
- Do students use their knowledge of part-to-whole relationships?
- Do students recognize that fractions refer to equal size parts?
- Can students use what they understand to move from part of a whole to part of a set?

Teacher Sound Bite

One of the most interesting aha moments for me as a third-grade teacher was when I gave the Crayon Count probe at the beginning of the school year. I actually agreed to it as part of a grade-span initiative in our school, and I considered it way too easy for my students. We begin our year with reviewing math ideas that include working with numbers in the thousands, so completing a simple question with a quantity of 12 seemed like a joke.

Well, my student's responses snapped me into a reality I had not anticipated. Many of my students were unable to make a numeric distinction between the three different color crayons as they related to the whole of 12. I was surprised with the difficulty they had in showing their thinking and ultimately providing an equation that represented their totals of each color crayon.

This experience reminded me how fragile student understanding is; merely because they work with large numbers doesn't mean they actually have understanding at a conceptual level for working with the ideas of fractions. I am much more explicit about my instruction now with my students, asking very specific questions that will uncover their understanding at the beginning, middle, and end of their third-grade year.

Additional References for Research and Teaching Implications

National Council of Teachers of Mathematics. (1993). *Research ideas for the classroom: Early childhood mathematics.* New York: Macmillan. (p. 105).

National Council of Teachers of Mathematics. (2000). *Principles and standards for school mathematics.* Reston, VA: Author. (pp. 82, 150).

National Council of Teachers of Mathematics. (2003). *Research companion to principles and standards for school mathematics.* Reston, VA: Author. (pp. 33–42).

Van de Walle, J. A. (2010). *Elementary and middle school mathematics* (7th ed.). Boston: Pearson. (p. 287).

Curriculum Topic Study and Uncovering Student Thinking

Crayon Count

Keeley, P., & Rose, C. (2006). *Mathematics curriculum topic study: Bridging the gap between standards and practice.* Thousand Oaks, CA: Corwin. (Fractions, p. 121; Algebraic Modeling, p. 135).

Related Elementary Probes:

Rose, C., Minton, L., & Arline, C. (2007). *Uncovering student thinking in mathematics: 25 formative assessment probes.* Thousand Oaks, CA: Corwin. (Fraction ID, p. 54; Fractional Parts, p. 49).

Student Responses to Crayon Count

Sample Responses: Incorrect

Student 1: "There are more purple because there are half of them. There are the same of orange and blue. There could be 10 of purple as the most and 1 and 1 of the other because 10 + 1 + 1 = 12."

Student 2: "I think there could be 8 purple and 2 orange and 2 blue because 8 + 2 = 10, and 10 + 2 = 12, and there were 12 crayons."

Sample Responses: Correct

Student 3: "I think that half of crayons are purple, so half of 12 is 6. Since half of crayons are purple, there are more of them than blue and just the same blue as orange. There are 6 purple, 3 blue, and 3 orange. I know because 6 + 3 + 3 = 12."

Student 4: "I started with 12 crayons. Half are purple, which is 6. That means there are 6 left. Half of 6 is 3, which is a quarter, so there are 3 left, so there are 3 of each orange and blue crayons."

PROBE 8A: VARIATION: CRAYON COUNT

There is a box of orange, blue, and purple crayons. Half ($\frac{1}{2}$) of the crayons are purple. Some crayons are orange, and a quarter ($\frac{1}{4}$) of the crayons are blue.

How many of the crayons are orange?

A) $\frac{3}{4}$

B) $\frac{2}{4}$

C) $\frac{1}{4}$

Explain your thinking:

_____

_____

_____

_____

_____

_____

_____

_____

_____

PROBE 9: IS $\frac{1}{4}$ OF THE WHOLE SHADED?

Decide if $\frac{1}{4}$ of the whole is shaded, and explain your thinking.

A)		Yes No	Explanation:
B)		Yes No	Explanation:
C)		Yes No	Explanation:
D)		Yes No	Explanation:
E)		Yes No	Explanation:
F)		Yes No	Explanation:
G)		Yes No	Explanation:
H)		Yes No	Explanation:
I)		Yes No	Explanation:
J)		Yes No	Explanation:

TEACHER NOTES: IS $\frac{1}{4}$ OF THE WHOLE SHADED?

Questioning for Student Understanding

Given a whole, are students able to indentify when $\frac{1}{4}$ of the whole is shaded?

K–2	3–5

Uncovering Understanding

Is $\frac{1}{4}$ of the Whole Shaded?: Content Standard: Number and Operations

Examining Student Work

Choosing incorrect examples or excluding correct examples may reveal *overgeneralizations* regarding part-whole relationships given an area model.

- The correct answers are A, B, C, F, H, and J. Students who choose each correct answer are using their knowledge of part-whole relationships using area models with a variety of shapes for the whole. These students understand the represented parts must be of equal size but not necessarily the same shape, orientation, or continuous. (See Student Response 1.)
- Inclusion of D: Students who choose D are typically counting the number of shaded parts. (See Student Responses 5.)
- Inclusion of E: Students who choose E are typically applying the ratio concept of one part shaded to four parts not shaded. (See Student Responses 6.)
- Inclusion of G: Students who choose G are typically confusing part-whole relationships with ordinal numbers (i.e., the fourth piece is shaded).
- Inclusion of I: Students who choose I typically believe that $\frac{1}{4}$ of the whole is shaded as long as one part is shaded out of four parts. These students are not yet paying attention to the size of the parts. (See Student Response 8.)
- Exclusion of C: Students who exclude this item typically are not familiar with the varying orientation of the parts of the whole. (See Student Response 4.)
- Exclusion of A and/or H and/or J: Students who exclude these items typically are not familiar with either the orientation of the shape or of the position of the shaded part. (See Student Response 2.)

- Exclusion of B and/or F: Students who exclude these items typically are not familiar with noncontinuous parts that together create $\frac{1}{4}$ of the whole or may view the shape as not being divided into four parts. (See Student Responses 3 and 7.)

Seeking Links to Cognitive Research

Of all the ways in which rational numbers can be interpreted and used, the most basic is the simplest—rational numbers are numbers. The fact is so fundamental that is easily overlooked. A rational number like $\frac{3}{4}$ is a single entity just as the number 5 is a single entity. Each rational number holds a unique place (or is a unique length) on the number line. Further, the way common fractions are written (e.g., $\frac{3}{4}$) does not help students see a rational number as a distinct number. Research has verified what many teachers have observed, that students continue to use properties they learned from operating with whole numbers even though many whole number properties do not apply to rational numbers. With common fractions, for example, students may reason that $\frac{1}{8}$ is larger than $\frac{1}{7}$ because 8 is larger than 7. Or they may believe that $\frac{3}{4}$ equals $\frac{4}{5}$ because in both fractions the difference between numerator and denominator is 1. (National Research Council [NRC], 2001. p. 235)

One of the primary characteristics of students' informal knowledge of fractions is that students' informal solutions involve separating units into parts and dealing with each part as though it represents a whole number, as opposed to dealing with each part as a fraction (Mack, 1990). For example, consider the following problem: If you have $\frac{5}{6}$ of a cake and I eat $\frac{2}{6}$ of the cake, how much of the cake do you have left? Students often refer to the fractions in the problem in terms of the "number or pieces" (e.g. five pieces or five pieces of six). However, the use of fraction names (e.g., five-sixths) refers to the fractions as specific parts of a whole. (NCTM, 2002b. p. 137)

Fractions make it possible to represent numbers between whole numbers. Fractions express numbers as an indicated division of two whole numbers. A fraction bar indicates the division. Some of the main ways that fractions are used in elementary school are as part of the whole, as quotient representations of ratios, as measures, as individual numbers on a number line, and in computation. (Bay Area Mathematics Taskforce, 1999. p. 59)

In grades K–2, in addition to work with whole numbers, young students should also have some experience with simple fractions through connections to everyday situations and meaningful problems, starting with the common fractions expressed in the language they bring to the classroom, such as "half." At this level it is more important for students to recognize when things are divided into equal parts than to focus on fraction notation. (NCTM, 2000. p. 82)

In grades 3–5, students should build their understanding of fractions as parts of a whole and as division. They will need to see and explore a variety of models of fractions, focusing primarily on familiar fractions such as halves, thirds, fourths, fifths, sixths eighths, and tenths. By using an area model in which part of a region is shaded, students can see how fractions are related to a unit whole, compare fractional parts of a whole, and fined equivalent fractions. (NCTM, 2000, p. 150)

*T*eaching Implications

In order to support a deeper understanding for students in elementary school in regard to fractions, specifically with identifying the size of fractional parts, the following are ideas and questions to consider in conjunction with the research.

Focus Through Instruction

- Use area models that allow students to see the part-to-whole relationship.
- Provide multiple experiences with various shapes when developing ideas related to part-to-whole relationships using area models.
- Provide multiple experiences with various orientations of shapes when developing ideas related to part-to-whole relationships using area models.
- Provide multiple experiences with various positions of the shaded pieces within shapes when developing ideas related to part-to-whole relationships using area models.
- Provide multiple experiences with various examples and nonexamples of given part-to-whole relationships (e.g., examples and nonexamples of $\frac{2}{3}$).
- Provide multiple experiences with various examples and nonexamples of given part-to-whole relationships with noncontinuous shaded parts.
- Use interact technology applets to connect from concrete materials to various representations. View example fraction applets at http://maine.edc.org/file.php/1/K6.html (see information about using interactive applets in Chapter 1, page 22).

Questions to Consider (when working with students as they grapple with the concepts of a fraction)

- How do students incorporate what they know about whole numbers when working with fractions?
- Do students understand that any shape can be defined as a whole and that the orientation of the shape does not matter?
- Do students understand that shaded regions do not have to be continuous?
- Do students understand the notion of equivalent pieces when determining part-to-whole relationships?

Teacher Sound Bite

This probe really brought out the varying levels of understanding of my third-grade class. I had several students at both ends of the spectrum . . . major misconceptions and no misconceptions and many in between. Almost all of them said yes to problem A, which let me know to incorporate additional activities where the parts are not the same. None of my students chose Item B, but I didn't feel it was the result of a misunderstanding but rather inexperience with the example. Knowing this allowed me to think of questions to push those students who had little difficulty with the probe.

Additional References for Research and Teaching Implications

National Council of Teachers of Mathematics. (1993). *Research ideas for the classroom: Early childhood mathematics.* New York: Macmillan. (pp. 133–134).

National Council of Teachers of Mathematics. (2002). *Putting research into practice in the elementary grades: Readings from journals of the NCTM.* Reston, VA. (pp. 59–70).

Van de Walle, J. A. (2007). *Elementary and middle school mathematics* (6th ed.). Boston: Pearson. (pp. 295–299).

Curriculum Topic Study and Uncovering Student Thinking

Is $\frac{1}{4}$ of the Whole Shaded?

Keeley, P., & Rose, C. (2006). *Mathematics curriculum topic study: Bridging the gap between standards and practice.* Thousand Oaks, CA: Corwin. (Fractions, p. 121).

Related Elementary Probes:

Rose, C., Minton, L., & Arline, C. (2007). *Uncovering student thinking in mathematics: 25 formative assessment probes.* Thousand Oaks, CA: Corwin. (Uncovering, p. 49; Fraction ID, p. 55).

Student Responses to Is $\frac{1}{4}$ of the Whole Shaded?

Sample Responses: Yes to A, B, C, F, H, and J

Student 1: "A, four pie pieces, and one is shaded; B, four squares, and one is shaded if I pretend the two shaded parts are moved together. C, Just flip the middle triangle around, and four triangles and one is shaded; F, move the four little grey squares to a corner, then there are four big squares, and one is shaded. H, just like before, four pie pieces, and one shaded. J, four rectangles, and 1 is shaded."

Sample Responses: Exclude A

Student 2: "The tire shape is close to $\frac{1}{4}$ of a pizza, so the piece of the pizza isn't."

Sample Responses: Exclude B

Student 3: "I drew in all the lines and then count and got 2 shaded out of 8."

Sample Responses: Exclude C

Student 4: "The pieces aren't all equal."

Sample Responses: Include D

Student 5: "4 are shaded."

Sample Responses: Include E

Student 6: "1 shaded, 4 not shaded."

Sample Responses: Exclude F

Student 7: "The unshaded pieces aren't same size and shape as the 1 shaded square piece."

Sample Responses: Include I

Student 8: "Just like I said in all the others that are right. There are 4 pieces and 1 of them is filled in."

PROBE 9A: VARIATION: IS $\frac{1}{4}$ OF THE WHOLE SHADED? CARD SORT

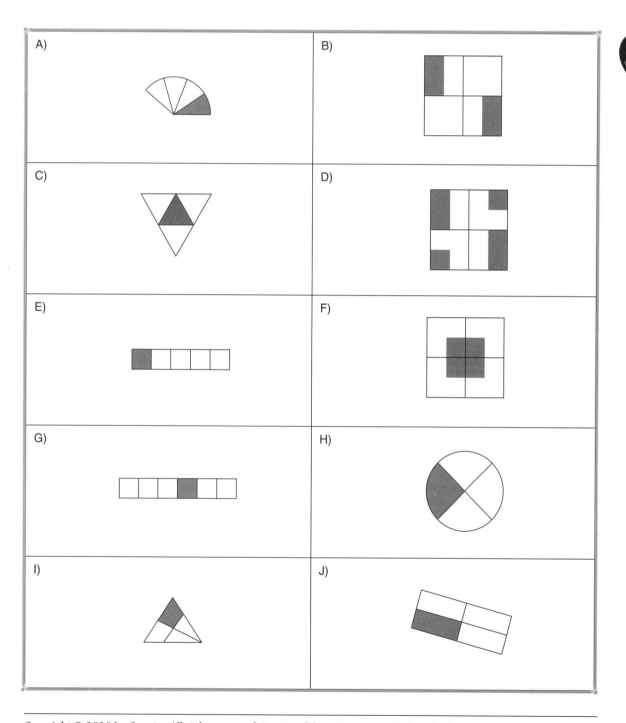

PROBE 9B: VARIATION: HOW MUCH IS SHADED?

What fraction of the shape is shaded?

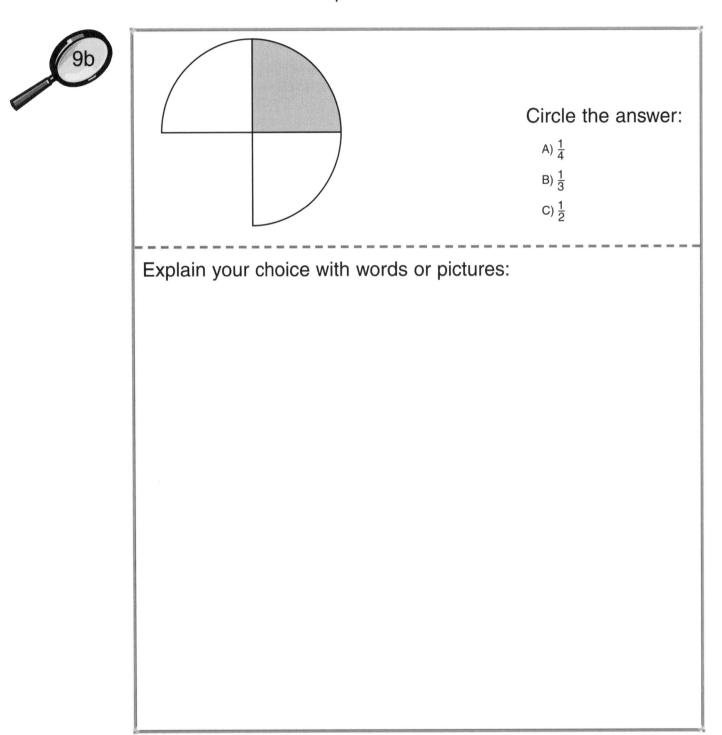

Circle the answer:

A) $\frac{1}{4}$

B) $\frac{1}{3}$

C) $\frac{1}{2}$

Explain your choice with words or pictures:

PROBE 10: GRANOLA BAR

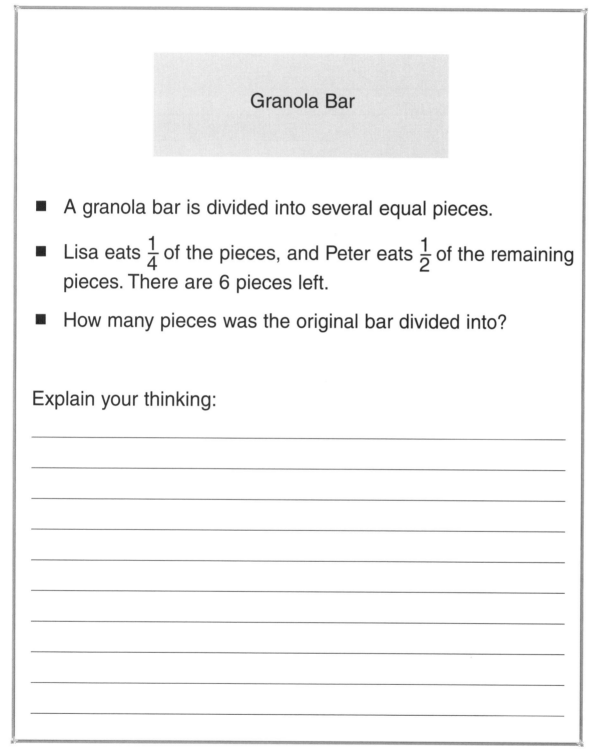

Granola Bar

- A granola bar is divided into several equal pieces.

- Lisa eats $\frac{1}{4}$ of the pieces, and Peter eats $\frac{1}{2}$ of the remaining pieces. There are 6 pieces left.

- How many pieces was the original bar divided into?

Explain your thinking:

TEACHER NOTES: GRANOLA BAR

Questioning for Student Understanding

Given an area model, are students able to define fractional parts of the whole?

K–2	3–5

Uncovering Understanding

Fractional Parts Content Standard: Number and Operation

Examining Student Work

Students' answers may reveal *overgeneralizations* regarding the application of whole-number properties to working with fractions.

- The correct response is 12 pieces. Students who answer 12 most likely consider the size of Lisa's pieces and Peter's pieces with the 6 remaining pieces as they relate to the whole. They are using their knowledge of part-whole relationships.
- Students who answered 16, 18, or 24: Students who answer 16, 18, or 24 considered the total number of pieces of the granola bar without relating the number to the size of the pieces. They most likely see all of the pieces as the total, much the same as in whole numbers, without considering the varying amounts indicated by Lisa and Peter.

Seeking Links to Cognitive Research

One of the primary characteristics of students' informal knowledge of fractions is that students' informal solutions involve separating units into parts and dealing with each part as though it represents a whole number, as opposed to dealing with each part as a fraction. (NCTM, 2002b, p. 137)

Fractions make it possible to represent numbers between whole numbers. Fractions express numbers as an indicated division of two whole numbers. A fraction bar indicates the division. Some of the main ways that fractions are used in elementary school are as part of the whole, as quotient representations of rations, as measures, as individual numbers on a number line, and in computation. (Bay Area Mathematics Task Force, 1999, p. 59)

Students develop an understanding of the meanings and uses of fractions to represent parts of a whole, parts of a set, or points or distances on a number line. They understand that the size of a fractional part is relative to the size of the whole, and they use fractions to represent numbers that are equal to, less than, or greater than 1. They solve problems that involve comparing and ordering fractions by using models, benchmark fractions, or common numerators or denominators. They understand and use models, including the number line, to identify equivalent fractions. (NCTM, 2006, p. 15)

In grades 3–5, students should build their understanding of fractions as parts of a whole and a s division. They will need to see and explore a variety of models of fractions, focusing primarily on familiar fractions such as halves, thirds, fourths, fifths, sixths, eighths, and tenths. By using an area model in which part of a region is shaded, students can see how fractions are related to a unit whole, compare fractional parts of a whole, and find equivalent fractions. They should develop strategies for ordering and comparing fractions, often using benchmarks such as $\frac{1}{2}$ and 1. (NCTM, 2000, p. 150)

<u>T</u>eaching Implications

In order to support a deeper understanding for students in elementary school in regard to fractions, specifically with using part-to-whole relationships to find the missing whole, the following are ideas and questions to consider in conjunction with the research.

Focus Through Instruction

- Students should make comparisons between numbers by using their understanding of equivalence.
- Students' interpretations of fractions should be varied to include region, set, and ratio situations.

- Use of an area model allows students to see the part-to-whole relationship.
- When working with a set interpretation, students must be able to understand that the "one" is a set and not a single object.
- Provide students with opportunity to record symbolically the number sentence and solution.
- Use interactive technology to have students practice measuring lengths
- An example site to consider for practicing these ideas is NCTM's Calculation Nation (2010) at http://calculationnation.nctm.org/Games/ (see information about using interactive applets in Chapter 1, page 22).

Questions to Consider
(when working with students as they
develop a deeper understanding of fractions)

- Do students use what they know about whole numbers when working with fractions?
- Can students make connections about relative size of fractions in relation to whole numbers?
- Can students apply appropriate methods when given various real-world contexts?
- Can students use mental images to consider relationships between sizes of fractions?
- Do students understand the inverse relationship between the number of pieces the whole unit is divided into and the size of the resulting individual pieces?

Teacher Sound Bite

I decided to use a mathematics probe at the end of a unit on fractions with my fifth-grade class. I wanted to know if my students had gained a better understanding of how to use what they understood about fractions and the part-to-whole relationship. I was pleased with the confidence my students showed in engaging with the task but a little disappointed in the results. Many students began to try to figure out the answer by using Lisa's and Peter's number of pieces without thinking about the fractional implications. The most common answer was 16. I am glad that I decided to use this probe in that way, so I could address this partial understanding with my students instead of assuming they had more knowledge then they really did.

Additional References for Research and Teaching Implications

Bay Area Mathematics Task Force. (1999). *A mathematics sourcebook for elementary and middle school teachers.* Novato, CA: Arena press. (p. 59).

National Council of Teachers of Mathematics. (2000). *Principles and standards for school mathematics.* Reston, VA: Author. (p. 150).

National Council of Teachers of Mathematics. (2002). *Putting research into practice in the elementary grades: Readings from journals of the NCTM.* Reston, VA: Author. (p. 137).

National Council of Teachers of Mathematics. (2006). *Curriculum focal points for prekindergarten through grade 8 mathematics: A quest for coherence.* Reston, VA: Author. (p. 15).

Curriculum Topic Study and Uncovering Student Thinking

Granola Bar

Keeley, P., & Rose, C. (2006). *Mathematics curriculum topic study: Bridging the gap between standards and practice.* Thousand Oaks, CA: Corwin. (Fractions, p. 121).

Related Elementary Probes: Crayon Count (page 85), Is $\frac{1}{4}$ of the Whole Shaded? (page 91)

Rose, C., Minton, L., & Arline, C. (2007). *Uncovering student thinking in mathematics: 25 formative assessment probes.* Thousand Oaks, CA: Corwin. (Fractional Parts, p. 49; Fraction ID, p. 54).

Student Responses to Granola Bar

Sample Responses: Incorrect

Student 1: "First, I wrote what $\frac{1}{4} + \frac{1}{2} = \frac{3}{4}$ and added 6 to it. Then, I picked 12 for a number because 6 + 6 = 12, so I did 12 + 6, which equals 16. As I thought about it, $\frac{1}{2}$ of 12 is 6, but that does not work, but 8 would work because ½ of it is 4. Anyway, I got 18 pieces."

Student 2: "I started by working backwards with the 6 pieces left over. I knew that Peter took ½ of the remaining pieces which would be another 6. Now, you have 12. $\frac{1}{4}$ would have to mean that if you took away $\frac{1}{4}$ of the pieces, it would be 12. 16 minus 4 is 12, and 16 minus $\frac{1}{4}$ = 16 minus 4, which equals 12. 16 is the amount of pieces in the beginning."

Sample Responses: Correct

Student 3: "I started by deciding that the 6 pieces left was an important clue. If 6 were left, then 6 were eaten because Peter ate $\frac{1}{2}$, which is 3 out of 6, and Lisa at a $\frac{1}{4}$, which is 3 because 3 × 4 = 12. So, the original bar had 12 pieces."

Student 4: "I drew a picture of a bar with $n + 6 = m$ as my equation. I then figured out that with $\frac{1}{2}$ the bar being 6, 12 was likely because $\frac{1}{4}$ of 12 is 3, and $\frac{1}{2}$ or 6 is 3, so 3 + 3 + 6 = 12; $n = 3$, and $m = 12$."

PROBE 11: IS IT EQUIVALENT?

Equivalent to $\frac{2}{5}$?		Explain why or why not:
A) 2.5	Yes No	
B) 25%	Yes No	
C) 0.4	Yes No	
D) 0.25	Yes No	
E) 40%	Yes No	
F) 2.5%	Yes No	
G) 0.04	Yes No	

TEACHER NOTES: IS IT EQUIVALENT?

Questioning for Student Understanding

Are students able to choose equivalent forms of a fraction?

K–2	3–5

Uncovering Understanding

Is It Equivalent? Content Standard: Number and Operations

Variation/Adaptation:

- Is It Equivalent? Card Sort

Examining Student Work

The distracters may reveal *common misunderstandings* regarding fraction and decimal-percentage conversions.

- The correct answers are C and E: Students who include each of the correct responses are able to find equivalent forms of $\frac{2}{5}$ and understand the relationship between fractions, decimals, and percentages. (See Student Responses 1 and 2.)
- Distracter A: Students who include A typically either divide 5 by 2 or replace the fraction bar with a decimal point. (See Student Response 3 and 4.)
- Distracter B: Students who include B are replacing the fraction bar with the percent symbol. (See Student Response 5.)
- Distracter F: Students who include F are replacing the fraction bar with both the decimal point and percent symbol. (See Student Response 6.)
- Distracter G: Students who include G typically understand to divide 2 by 5 but are not sure of the decimal point placement. (See Student Response 7.)

Seeking Links to Cognitive Research

Elementary-school students may have difficulties perceiving a fraction as a single quantity (Sowder, 1988), but rather see it as a pair of whole numbers. An intuitive basis for developing the concept of fractional number is provided by partitioning (Kieren, 1992) and by seeing fractions as multiples of basic units—for example, $\frac{3}{4}$ is $\frac{1}{4}$ and $\frac{1}{4}$ and $\frac{1}{4}$ rather than 3 of 4 parts (Behr et al., 1983). (American Association for the Advancement of Science [AAAS], 1993, p. 350)

By studying fractions, decimals, and percents simultaneously, students can learn to move among equivalent forms, choosing and using an appropriate and convenient form to solve problems and express quantities. (NCTM, 2000, p.150)

One of the first challenges facing a young student is that a rational number can take multiple forms. The learner must be able to efficiently and effectively move between these forms [i.e., fractions, decimals, and percents]. This flow between representations does not come easy. (NRC, 2005, p. 213)

To connect the two numeration systems, fractions and decimals, students should make concept-oriented translations; that is, translations based on understanding rather than a rule or algorithm. (Van de Walle, 2007, p. 337)

We should extend familiarity with simple fractions [especially halves, thirds, fourths, fifths, and eighths] to the same concepts expressed as decimals. One way to do this is to have students translate familiar fractions to decimals by means of a base-ten model. (Van de Walle, 2007, p. 338)

Students should use base-ten models for percents in much the same way as for decimals. (Van de Walle, 2007, p. 344)

Teaching Implications

To support a deeper understanding for students in intermediate grades in regards to number and operations, the following are ideas and questions to consider in conjunction with the research.

Focus Through Instruction

- Use a variety of models and representations, such as fraction strips, number lines, grid paper, area models (rectangles and circles), and symbolic representations.
- Connections between fractions and decimals and fractions and percentages should be explicit.
- A solid foundational understanding of rational numbers is important for students to be able to convert between multiple representations.
- Use interactive technology applets to bridge between concrete and abstract concepts. See the example Base Block virtual manipulative at http://nlvm.usu.edu/en/nav/category_g_3_t_1.html. (See information about using interactive applets in Chapter 1, page 22.)

Questions to Consider (when working with students as they grapple with the concepts of a fraction)

- Do students understand that one meaning of the fraction bar is division?
- Are students able to represent fractions in a variety of ways, including use of area and linear models, as parts of sets, and symbolically (all of which contribute necessary foundational knowledge)?

- Are students able to represent percents, fractions, and decimals in ways that make for easy comparison?
- Do students understand the relationship between fractions, decimals, and percentages?
- Do students understand that the numerator and denominator of a fraction are related by multiplication rather than addition?
- Do students understand that a fraction (and all equivalent fractions) can only be equivalent to one decimal representation and one percent representation?

Teacher Sound Bite

I was taken aback a bit to see the number of students who thought a fraction can be equivalent to more than one decimal or percent. About a third of my students circled yes for both 2.5 and .4, and about a quarter of them chose 25%, 2.5%, and 40%. Because many of these students can successfully convert a fraction to a decimal or percent when asked, I would not have known they had this other issues with also thinking additional decimals and percents were also equivalent. Once I realized that students overgeneralized the idea that there are an infinite number of equivalent fractions to the idea that a fraction can convert to multiple decimals and percent forms, I was able to address the misconceptions head on in classroom discussions.

Additional References for Research and Teaching Implications

American Association for the Advancement of Science. (1993). *Benchmarks for science literacy.* New York: Oxford University Press. (pp. 350, 358–359).

National Council of Teachers of Mathematics. (1993). *Research ideas for the classroom: Early childhood mathematics.* New York: Macmillan. (pp. 224–231).

National Research Council. (2001). *Adding it up: Helping children learn mathematics.* Washington, DC: National Academy Press. (p. 235).

Van de Walle, J. A. (2007). *Elementary and middle school mathematics* (6th ed.). Boston: Pearson. (pp. 337–341).

Curriculum Topic Study and Uncovering Student Thinking

Is It Equivalent?

Keeley, P., & Rose, C. (2006). *Mathematics curriculum topic study: Bridging the gap between standards and practice.* Thousand Oaks, CA: Corwin. (Fractions, Decimals, and Percentages, p. 122).

Related Elementary Probe:

Rose, C., & Arline, C. (2009). *Uncovering student thinking in mathematics, grades 6–12: 30 formative assessment probes for the secondary classroom.* Thousand Oaks, CA: Corwin. (Variation: Is It Equivalent? p. 45).

Student Responses to Is It Equivalent?

Sample Responses: C and E

Student 1: "I know $\frac{1}{5}$ is 20%, so $\frac{2}{5}$ must be 40%. Same thing with decimals. I know $\frac{1}{5}$ is the same as .2, so $\frac{2}{5}$ is .4."

Student 2: "$\frac{2}{5}$ is $\frac{4}{10}$, and $\frac{4}{10}$ is 0.4, then from 0.4 you can get to 40% cuz .4 is .40, and .40 is 40 hundredths."

Sample Responses: A

Student 3: "2.5 is the same thing as $\frac{2}{5}$."

Student 4: "It's like sharing 5 donuts with 2 kids. They each get 2 and a half."

Sample Responses: B

Student 5: "It is the same numbers but just with a % sign instead of a fraction."

Sample Responses: D

Student 6: "Because it uses the same numbers."

Sample Responses: F

Student 7: "Just know that % and . are related to fractions."

Sample Responses: G

Student 8: "Take 2 and divide by 5. Since it doesn't go in, add extra zeros, then divide then put decimal point back in."

PROBE 11A: VARIATION: IS IT EQUIVALENT? CARD SORT

A) **2.5**	E) **40%**
B) **25%**	F) **2.5%**
C) **0.4**	G) **0.04**
D) **0.25**	

11a

PROBE 12: IS IT SIMPLIFIED?

Decide whether or not each example shows a correct use of "canceling zeros" to simplify a fraction.

Example 1 $\dfrac{50}{70}$

$\dfrac{5\cancel{0}}{7\cancel{0}}$ **So**

$\dfrac{50}{70} = \dfrac{5}{7}$

Example 1:   Correct   Incorrect

Explain your choice:

Example 2 $\dfrac{501}{702}$

$\dfrac{5\cancel{0}1}{7\cancel{0}2}$ **So**

$\dfrac{501}{702} = \dfrac{51}{72}$

Example 2:   Correct   Incorrect

Explain your choice:

Example 3 $\dfrac{30}{705}$

$\dfrac{3\cancel{0}}{7\cancel{0}5}$ **So**

$\dfrac{30}{705} = \dfrac{3}{75}$

Example 3:   Correct   Incorrect

Explain your choice:

TEACHER NOTES: IS IT SIMPLIFIED?

$\underline{Q}$uestioning for Student Understanding

Do students use the "canceling of zeros" shortcut appropriately?

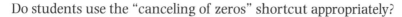

K–2	3–5

$\underline{U}$ncovering Understanding

Is It Simplified? Content Standard: Number and Operations

$\underline{E}$xamining Student Work

The distracters may reveal lack of *conceptual understanding* of the use of the common shortcut of "canceling" the zeros to simplify a fraction. Often, students' lack of understanding of rational numbers hinders their ability to justify by comparing in cases of incorrect canceling of digits.

- The correct answers are Example 1, correct; Example 2, incorrect; and Example 3, incorrect: Students who choose these responses are correctly applying the canceling zeros rule as a specific case by showing understanding of dividing the numerator and denominator by a common factor, which in these special cases are a multiple of 10. (See Student Response 1.)
- Distracter Example 2: Students who inappropriately choose "correct" tend to view zero as "only" a place holder, and although they often recognize the need for the zeros to be in the same position in both numbers, there is a lack of understanding of the need for the numbers to also be multiples of 10. (See Student Responses 2 and 3.)
- Distracter Example 3: Students who inappropriately choose "correct" *overgeneralize* the rule to situations where both the numerator and denominator contain a zero regardless of the place in which the zero is located. Typically, these students are incorrectly applying a procedure with no attempt at justifying by comparing the resulting fraction to the original. (See Student Responses 4 and 5.)

$\underline{S}$eeking Links to Cognitive Research

The role of zero as a placeholder in the symbolic representation of number is frequently documented as problematic for children learning to read and write numbers in conventional formats (Wheeler & Feghali, 1983, p. 147). (AAAS, 1993, p. 350)

Care must be taken in sequencing the development of the number names and the number symbols to allow an appreciation of the lack of particular units and the use of zero to signify that there is "none of that place" (Booker et al., 1997). (Anthony & Walshaw, 2004, p. 41)

Zero and, to a lesser extent, 1 as factors often cause difficulty for children. (Van de Walle, 2007, p. 157)

There is value in exposing students early to products involving multiples of 10 and 100. . . . Be aware of students who simply tack on zeros without understanding why. (Van de Walle, 2007, p. 231)

<u>T</u>eaching Implications

To support a deeper understanding for students in elementary grades in regards to number and operations, the following are ideas and questions to consider in conjunction with the research.

Focus Through Instruction

- Provide instruction that is concrete and process oriented rather than abstract and procedure oriented.
- Help students generate the rules of multiplying and dividing by multiples of 10 from their experiences with manipulatives.
- Experiences with true-false number sentences provide children with experiences to apply operations with multiples of 10.
- Using a common-factor approach can help students build conceptual understanding.

Questions to Consider (when working with students as they grapple with ideas related to the concept of fraction)

- Do students understand simplifying as dividing by a common factor?
- Are students able to compare fractions?
- Are students developing their own generalized rules based on conceptual understanding and pattern recognition?
- Are students able to justify appropriate use of the generated rule?
- When giving explanations, such as canceling the zeros or moving the decimal point, can students answer the question, When and why does that procedure result in a correct solution?

Teacher Sound Bite

I realize the importance of my own use of correct mathematical language in order to not promote the incorrect use of short cuts. Most of the math I learned in my own K–12 experience was based on carrying out rules and procedures without having an understanding of why I was doing each step. One of my goals related to math terminology is to never use the term "cancel the zeros" but instead to refer to simplifying by dividing by whatever multiple of 10 is appropriate for the problem.

Additional References for Research and Teaching Implications

Anthony, G. J., & Walshaw, M. A. (2004). Zero: A "none" number? *Teaching Children Mathematics, 11*(1), 38–42.

Bay Area Mathematics Task Force. (1999). *A mathematics sourcebook for elementary and middle school teachers.* Novato, CA: Arena Press. (pp. 59–70).

National Council of Teachers of Mathematics. (1993). *Research ideas for the classroom: Early childhood mathematics.* New York: Macmillan. (pp. 118–134).

Curriculum Topic Study and Uncovering Student Thinking

Is It Simplified?

Keeley, P., & Rose, C. (2006). *Mathematics curriculum topic study: Bridging the gap between standards and practice.* Thousand Oaks, CA: Corwin. (Fractions, p. 121).

Rose, C., & Arline, C. (2009). *Uncovering student thinking in mathematics, grades 6–12: 30 formative assessment probes for the secondary classroom.* Thousand Oaks, CA: Corwin. Variation: Is It Simplified? p. 61).

Student Responses to Is It Simplified?

Sample Responses: A and F

Student 1: "Correct because divide by 10."

Sample Responses: Inclusion of B

Student 2: "Correct. Don't need the zeros."

Student 3: "Incorrect. It only works sometimes, like $\frac{100}{200}$ is ok, but $\frac{205}{250}$ not ok because anyone can tell $\frac{25}{50}$ is $\frac{1}{2}$ but $\frac{205}{250}$ is not $\frac{1}{2}$."

Sample Responses: Inclusion of C

Student 4: "Correct. Just like I said before, you don't need the zeros."

Student 5: "Incorrect. The zeros have to be in the same column before you can cancel them."

5

Structure of Number Probes

Computation and Estimation

Figure 5.1   Chapter 5 Probes

Grade-Span Bar Key							
	Target for Instruction Depending on Local Standards						
	Prerequisite Concept and Field Testing Indicate Students May Have Difficulty						

Question	Probe	Grade Span					
Structure of Number: Computation and Estimation		K	1	2	3	4	5
Do students use the structure of ten when combining collections?	How Many Dots? (page 116)	K	1–2				
When considering the whole and two parts, can students identify all possible part-part-whole combinations?	Play Ball (page 122)		1–2		3		
When adding, can students apply and understand a variety of different strategies?	What's Your Addition Strategy? (page 128)			2–3		4–5	
When subtracting, can students apply and understand a variety of different strategies?	What's Your Subtraction Strategy? (page 136)			2–3		4–5	
Are students flexible in using strategies for solving various multiplication problems?	What's Your Multiplication Strategy? (page 144)					4–5	
Are students flexible in using strategies for solving division problems?	What's Your Division Strategy? (page 150)					4–5	
Do students understand there are multiple methods of estimating the sum of three 2-digit numbers?	Is It an Estimate? (page 155)					4–5	
Can students use estimation to choose the closest benchmark to an addition problem involving fractions?	What Is Your Estimate? (page 160)					4–5	

PROBE 13: HOW MANY DOTS?

(Student Interview Task)

This task is an interview conducted with one student at a time. A cover screen is required to screen ten-frames B and C as the child answers Question 1. Ten-frame B is then revealed, and the student answers Question 2. The screen is removed, revealing all three ten frames, and the student answers Question 3.

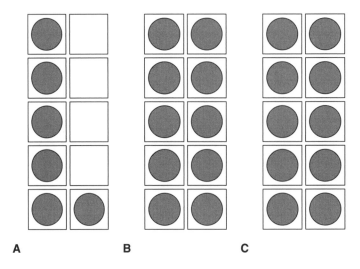

Question 1:

Teacher asks, "How many dots are in this ten frame?" Student response _____

Circle student strategy: Count All   5 + 1   10 – 4   Other _____

Question 2:

Teacher asks, "How many dots are there now?" Student response _____

Circle student strategy: Count All   6 + 10   20 – 4   Other _____

Question 3:

Teacher asks, "How many dots are there now?" Student response _____

Circle student strategy: Count All   16 + 10   30 – 4   Other _____

**If not obvious, ask student how he or she figured it out after each question.

*You are looking for students to use the structure of ten as they calculate how many dots.

TEACHER NOTES: HOW MANY DOTS?

Questioning for Student Understanding

Do students use the structure of ten when combining collections?

K–2	3–5

Uncovering Understanding

Structure of Ten: Whole Numbers Content Standard: Number and Operation

Variation/Adaptation:

- Blank Template to Create Own Version

Examining Student Work

Students' answers may reveal *common errors* due to a lack of conceptual knowledge regarding the idea of cardinality and the structure of ten. A common misconception may be revealed as students count the dots by ones without considering using the structure of the ten frame. Do they use the structure to add a 10 and/or recount the previous quantity each time, even though it has already been counted previously?

- Strategy—Count All: Students who count all 6 dots for frame A and then for frame B count all 6 again and then the next 10 are not using the structure of ten and rely on "counting all" as their primary addition strategy.
- Strategy—Use 10: Students who count all 6 dots for frame A and then for frame B hold the 6 and count on by adding 10 are using their understanding of a full frame being a quantity of 10. Continuing to add a 10 for frame C by saying, "16 + 10 is 26," further demonstrates efficiency in using the unit of ten to join multiple sets.

Seeking Links to Cognitive Research

During preschool and elementary school years, children develop meanings for number words in which sequence, count, and cardinal meanings of number words become increasingly integrated. (American Association for the Advancement of Science [AAAS], 1993, p. 350)

Children should learn that the last number named represents the last object as well as the total number of objects in the collection. Concrete models can help students represent numbers and develop number sense; they can also help bring meaning to students' use of written symbols and can be useful in building place-value concepts. (National Council of Teachers of Mathematics [NCTM], 2000, pp. 79–80)

Children use numbers, including written numerals, to represent quantities and to solve quantitative problems, such as counting objects in a set, creating a set with a given number of objects, comparing and ordering sets or numerals by using both cardinal and ordinal meanings, and modeling simple joining and separating situations with objects. (NCTM, 2006, p. 12)

The ten-frame is a useful devise for helping children visualize sums between 6 and 10 in terms of patterns based on 5. (NCTM, 1993a, p. 87)

The application of numerical understanding is carried out by the development of strategies. The significance of counting cannot be underscored enough in terms of what it means for students developmentally. The strategies students construct in those early years become their foundation on which to balance subsequent skills and concepts. (Minton, 2007, p. 57)

Making a transition from viewing "ten" as simply the accumulation of 10 ones to seeing it both as 10 ones *and* as 1 ten is an important first step for students toward understanding the structure of the base-ten number. (NCTM, 2003, p. 33)

Understanding number requires much more than verbal counting. It also includes the ability to determine the total number of objects and reasoning about that numerosity using number relationships. Numerosity and reasoning are influenced by the size of the numbers and the ability to think using groups. Since numbers are used in a variety of ways and with a variety of symbols, context and symbols are added influence on children's understanding of number. (NCTM, 1993a, p. 44)

Children develop an understanding of the base-ten numeration system and place-value concepts (at least to 1000). Their understanding of base-ten numeration includes ideas of counting in units and multiples of hundreds, tens, and ones, as well as a grasp of number relationships, which they demonstrate in a variety of ways, including comparing and ordering numbers. They understand multi-digit numbers in terms of place value, recognizing that place-value notation is a shorthand for the sums of multiples of powers of 10 (e.g., 853 as 8 hundreds + 5 tens + 3 ones). (NCTM, 2006, p. 14)

_T_eaching Implications

To support an initial understanding for early elementary students in regard to "trusting the count" (cardinality) of a set based on the structure of 5 and 10, the following ideas and questions to consider in conjunction with the research.

Focus Through Instruction

- Pay careful attention to children's ability to count accurately.
- Provide opportunities for them to confirm the concept of cardinality by following up their initial response to a total by asking them again, "So, how many was that?"
- Notice what students do computationally that reveals what they understand about using tens.
- Give practice in instant recognition of small quantities (i.e., subitizing).
- Emphasize place-value ideas by asking questions that support the base-ten number system (e.g., "What is 10 more? 10 less?").
- Use of calculators, ten frames, linking cubes, and base-10 blocks can help to develop and reinforce using tens.
- Use interactive technology to support student understanding. View an example activity from NCTM's Illuminations (2000–2010) site at http://illuminations.nctm.org/ActivityDetail.aspx?ID=74 and http://illuminations.nctm.org/ActivityDetail.aspx?ID=75 (see information about using interactive applets in Chapter 1, page 22).

Questions to Consider
(when working with students as
they develop an understanding of 10)

- When counting collections, do students recognize that the last number said represents the total for the collection?
- When counting two collections, do students continue to count all items regardless of the size of the two collections?
- Do student strategies vary based on what they understand about the numbers involved in a problem, or do they apply the same strategy regardless of the numbers?
- Do students use a systematic way to join two or more collections?
- Do students have a mental model of quantities to five?

Teacher Sound Bite

I watched my first graders count, read, identify, and join two-digit numbers through 50, so I believed that they understood the "ten-ness" of our number system. We use popsicle sticks during calendar time and dimes and pennies for days in school, so I was very surprised when I used How Many Dots? how many of my students counted by ones to join the collections rather than use what they knew about 10. I have used many models and representations and mistook the rote responses to suggest that they understood 10 and, what is more important, that they were using it to add. I am now adding reasoning questions to my math routines to ensure that I am providing students with conceptual and procedural levels of addition.

Curriculum Topic Study and Uncovering Student Thinking

How Many Dots?

Keeley, P., & Rose, C. (2006). *Mathematics curriculum topic study: Bridging the gap between standards and practice.* Thousand Oaks, CA: Corwin. (Addition and Subtraction, p. 111; Counting, p. 116; Elementary Number Theory, p. 119; Facts, p. 120; Number Sense, p. 127).

Rose, C., Minton, L., & Arline, C. (2007). *Uncovering student thinking in mathematics: 25 formative assessment probes.* Thousand Oaks, CA: Corwin. (Is One Group More? p. 32; Ducks in a Row, p. 37).

Additional References for Research and Teaching Implications

Minton, L. (2007). *What if your ABCs were your 123s?: Building connections between literacy and numeracy.* Thousand Oaks, CA: Corwin. (p. 57).

National Council of Teachers of Mathematics. (1993). *Research ideas for the classroom: Early childhood mathematics.* Reston, VA: Author. (pp. 44, 87).

National Council of Teachers of Mathematics. (2000). *Principles and standards for school mathematics.* Reston, VA: Author. (p. 80–84).

National Council of Teachers of Mathematics. (2003). *Research companion to principles and standards for school mathematics.* Reston, VA: Author. (pp. 33–42).

National Council of Teachers of Mathematics. (2006). *Curriculum focal points for prekindergarten through grade 8 mathematics: A quest for coherence.* Reston, VA: Author. (p. 14).

Student Responses to How Many Dots?

Sample Responses

Student 1: Question 1: "I think there are 6 because I counted them."

Question 2: "One, two, three . . . six." Then, "One, two, three . . ." Counted all again to get 16.

Question 3: "One, two, three . . . six." Then, "One, two, three . . . 10, Oh, that's 10 too." So, counted all again to get 26.

Student 2: Question 1: "Six, because I counted all of them and got six."

Question 2: "16, because that is a 10 and 6 + 10 = 16."

Question 3: "26, I don't need to count because it is just 10 more, so that is just like skip counting."

PROBE 13A: VARIATION: HOW MANY COUNTERS?

PROBE 14: PLAY BALL

There are *some* baseballs and *some* soccer balls loose on the gym floor. The teacher asks you to pick up all 8 balls to put them away.

How many baseballs and how many soccer balls *could* there be on the gym floor?

- -

Use pictures, numbers, and/or words to show how many of each you *could* pick up.

TEACHER NOTES: PLAY BALL

<u>Q</u>uestioning for Student Understanding

When considering the whole and two parts, can students identify all possible part-part-whole combinations?

K–2	3–5

<u>U</u>ncovering Understanding

Part-Part-Whole Content Standard: Number and Operation

Variation/Adaptation:

- Variation: Play Ball

<u>E</u>xamining Student Work

Students' answers may reveal *partial understanding* of the concept of part-part-whole relationships. This is a big idea in the development of number sense that leads to understanding the properties of addition and subtraction. Students need to understand that a number is made up of two *or more* parts, and those parts can be put together to make the whole. It is crucial that students in the primary grades come to understand the part-part-whole relationship.

- The correct response is 4 + 4, 5 + 3, 6 + 2, 7 + 1, and the commutative pairs: Students who included these combinations as possibilities looked to the context and demonstrated the ability to break 8 down into two parts that represented *some baseballs and some soccer balls.* (Note: 8 + 0 is an incorrect response; although it is such a relationship, it does not apply to the context of this probe.)
- Single relationship response: Students who provided a single response to the question, such as there are four baseballs and four soccer balls, or five baseballs and three soccer balls, have a partial understanding of part-part-whole yet are not generalizing the concept to include other representations for 8.

<u>S</u>eeking Links to Cognitive Research

As students work with numbers, they gradually develop flexibility in thinking about number, which is a hallmark of number sense. . . . Number sense develops as students understand the size of numbers, develop multiple ways of thinking about and representing numbers, use numbers as referents, and develop accurate perceptions about the effects of operations on numbers. (NCTM, 2000, p. 80)

Probably the major conceptual achievement of the early school years is the interpretation of number in terms of part and whole relationships. With the application of Part-Whole schema to quantity, it becomes possible for children to think about numbers as compositions of other numbers. This enrichment of number understanding permits forms of mathematical problem solving and interpretation that are not available to younger children. (Resnick & Omanson, 1987, p. 114)

Children use their understanding of addition to develop quick recall of basic addition facts and related subtraction facts. They solve arithmetic problems by applying their understanding of models of addition and subtraction, relationships and properties of number and properties of addition. (NCTM, 2006, p. 14)

Number relationships provide the foundation for strategies that help students remember basic facts. For example, knowing how numbers are related to 5 and 10 helps students master facts such as 3 + 5 and 8 + 6. (Van de Walle, 2010, p. 167)

Children use mathematical reasoning, including ideas such as commutativity and associativity and beginning ideas of tens and ones, to solve two-digit addition and subtraction problems with strategies that they understand and can explain. They solve both routine and non-routine problems. (NCTM, 2006, p. 13)

_T_eaching Implications

To support an initial understanding for early elementary students in regard to the concept of part-part-whole relationships or decomposing numbers, the following are ideas and questions to consider in conjunction with the research.

Focus Through Instruction

- Encourage students to take apart a number into two or more parts in more than one way.
- Use manipulatives to reinforce what the number looks like as students then break it apart into smaller parts.
- Focus on a quantity in terms of the whole and its parts.
- Focus on encouraging a search for and discussion of patterns and relationships.

- Basic facts are developmental, and students should get to a "just knowing" of recall.
- Use models and graphic organizers that show the relationship of part-part-whole.
- Use interactive technology to support student understanding. View an example activity from NCTM's Illuminations (2000–2010) site at http://illuminations.nctm.org/LessonDetail.aspx?id=U147 and http://illuminations.nctm.org/LessonDetail.aspx?ID=U153 (see information about using interactive applets in Chapter 1, page 22).

Questions to Consider (when working with students as they develop an understanding of part-part-whole relationships)

- Can students move between the parts of a number and the whole fluently?
- Do students recognize the relationship between a whole and more than two parts?
- Are students able to compose and decompose numbers to include multiple combinations of the whole?
- Can students share their reasoning for their strategies?
- Do students make connections between the properties of addition and addition facts?

Teacher Sound Bite

After giving the Play Ball probe, it became very clear to me that my students were able to recall number-fact combinations, especially the doubles, with great enthusiasm and accuracy when asked to respond to a particular prompt. What was missing from my instruction, as indicated by the probe, was providing students with the opportunity to consider that multiple combinations could be used within a context to answer a question. Most of my students answered with 4 soccer balls and 4 baseballs with the reasoning that 4 + 4 = 8. It did not occur to them that there were other possibilities for the combinations of balls. As a result, I am adding a scenario similar to Play Ball each week during morning messages to provide my students with the opportunity to move past their initial response and determine if there are other possibilities. Sharing their thinking allows the students an access point and encourages them to think more deeply to decide if they have found all the possible combinations. This is an important idea as students begin to use larger numbers and think about how to compose and decompose them efficiently. If they develop one combination for a particular number, they will have great difficulty generalizing their fact knowledge to larger numbers, which is likely to impede their computational fluency.

**Curriculum Topic Study
and Uncovering
Student Thinking**

Play Ball

Keeley, P., & Rose, C. (2006). *Mathematics curriculum topic study: Bridging the gap between standards and practice.* Thousand Oaks, CA: Corwin. (Addition and Subtraction, p. 111; Computation and Operations, p. 115; Numbers and Number Systems, p. 128).

Related Elementary Probes: Variation: Play Ball (page 127)

Rose, C., Minton, L., & Arline, C. (2007). *Uncovering student thinking in mathematics: 25 formative assessment probes.* Thousand Oaks, CA: Corwin.

Additional References for Research and Teaching Implications

National Council of Teachers of Mathematics. (1993). *Research ideas for the classroom: Early childhood mathematics.* New York: Macmillan. (pp. 44, 87).

National Council of Teachers of Mathematics. (2000). *Principles and standards for school mathematics.* Reston, VA: Author. (pp. 80–84).

National Council of Teachers of Mathematics. (2006). *Curriculum focal points for prekindergarten through grade 8 mathematics: A quest for coherence.* Reston, VA: Author. (p. 14).

Van de Walle, J. A. (2010). *Elementary and middle school mathematics* (7th ed.). Boston: Pearson. (p. 167).

Student Responses to Play Ball

Sample Responses: Single Relationship

Student 1: "You can have 4 baseballs and 4 soccer balls because 4 + 4 = 8. I know I am right because 4 x 2 = 8."

Student 2: "I think you can have 7 baseballs and 1 soccer ball or 7 soccer balls and 1 baseball, but either one would equal 8. I know because 8 is just one more than 7."

Sample Responses: Multiple Relationship

Student 3: "You can have lots of different ways if you count up and then down. 7 + 1, 6 + 2, 5 + 3, 4 + 4, 3 + 5, 2 + 6, 1 + 7. I know this because we talk about showing all ways in math class. Pairs that equal 8 except for 8 + 0 because the question says there are some of each. 7 + 1, 6 + 2, 5 + 3, 4 + 4, and turnaround facts."

PROBE 14A: VARIATION: PLAY BALL

There are *some* baseballs and *some* soccer balls loose on the gym floor. The teacher asks you to pick up all 10 balls to put them away.

How many baseballs and how many soccer balls *could* there be on the gym floor?

"There are 10 balls, so 5 baseball and 5 soccer balls. Because 5 + 5 = 10.

"I think you can do 5 + 5 = 10, but there are lots more other ways too."

Who do you think is correct, Amelia or Elias? Explain your thinking:

_____

_____

_____

_____

_____

PROBE 15: WHAT'S YOUR ADDITION STRATEGY?

Sam, Julie, Pete, and Lisa each added the numbers 34 and 56.

1. *Circle* the method that most closely matches how you solved the problem.

2. *Explain* whether *each* method makes sense mathematically.

A) Sam's Method	
3^{1}4 + 5 6 ―――― 90	

B) Julie's Method	
34 + 56 ―――― 80 10 ―――― 90	

C) Pete's Method	
34 + 56 +6 −6 40 + 50 = 90	

D) Lisa's Method	
56 + 34 56 + 20 = 76 76 + 10 = 86 86 + 4 = 90	

TEACHER NOTES: WHAT'S YOUR ADDITION STRATEGY?

Questioning for Student Understanding

When adding, can students apply and understand a variety of different strategies?

K–2	3–5

Uncovering Understanding

Addition Strategies: Whole Numbers Content Standard: Number and Operation

Important Note: Prior to giving students the probe, ask them to individually solve the indicated problem. If time allows, ask them to solve the problem in at least two or three different ways.

Variations/Adaptations:
- Addition Strategies: Three-Digit Numbers
- Addition Strategies: Decimals

Examining Student Work

Student answers may reveal *misunderstandings* regarding methods of addition, including lack of *conceptual understanding* of the properties of numbers. Responses also may reveal a common misconception that there is only one correct algorithm for each operation or that, once comfortable with a method, there is no need to understand other methods.

- Sam's Method: This method is usually recognized by third- to fifth-grade students, although in some situations students may not have been introduced to this standard U.S. algorithm. Those who have no experience with the method show a lack of procedural understanding of the algorithm and typically indicate the method "does not make sense because 1 + 34 + 56 is 91, not 90." Those students who recognize the algorithm often do not demonstrate place-value understanding. (See Student Responses 1 and 2.)
- Julie's Method: This method is often recognized by students who have experience with multiple algorithms as well as those who were taught only the traditional algorithm. These latter students often apply variations of using an expanded notation form of the numbers. (See Student Response 3.)

- Pete's Method: This strategy is often the least recognized by students in terms of generalizing a method of adding and subtracting like amounts from the numbers to keep a constant total. (See Student Response 4.)
- Lisa's Method: Students who use this strategy hold the first number constant and break the addend into place-value parts, adding on one part at a time. Students who have experience using an open number line are typically able to mathematically explain this method of addition. (See Student Response 5.)

<u>S</u>eeking Links to Cognitive Research

Student errors when operating on whole number suggest students interpret and treat multi-digit numbers as single-digit numbers placed adjacent to each other, rather than using place-value meanings for digits in different positions. (AAAS, 1993, p. 358)

The written place-value system is a very efficient system that lets people write very large numbers. Yet it is very abstract and can be misleading: The digits in every place look the same. To understand the meaning of the digits in the various places, children need experience with some kind of *size-quantity supports* (e.g., objects or drawings) that show tens to be collections of 10 ones and show hundreds to be simultaneously 10 tens and 100 ones, and so on. (NCTM, 2003, p. 78)

Students can use roughly three classes of effective methods for multi-digit addition and subtraction, although some methods are mixtures. *Counting list methods* are extensions of the single-digit counting methods. Children initially may count large numbers by ones, but these unitary methods are highly inaccurate and are not effective. All children need to be helped as rapidly as possible to develop prerequisites for methods using tens. These methods generalize readily to counting on or up by hundreds but become unwieldy for larger numbers. In *decomposing methods,* children decompose numbers so that they can add or subtract the like units (e.g., add tens to tens, ones to ones, hundreds to hundreds). These methods generalize easily to very large numbers. *Recomposing methods* are like the make-a-ten or doubles methods. The solver changes both numbers by giving some amount of one number to another number (i.e., in adding) or by changing both numbers equivalently to maintain the same difference (i.e., in subtracting). (NCTM, 2003, p. 79)

When students merely memorize procedures, they may fail to understand the deeper ideas. When subtracting, for example, many children subtract the smaller number from the larger in each column, no matter where it is. (National Research Council [NRC], 2002, p. 13)

By the end of the 3–5 grade span, students should be computing fluently with whole numbers. Computational fluency refers to having efficient and accurate methods for computing. Students exhibit computational fluency when they demonstrate flexibility in the computational methods they choose, understand and can explain these methods, and produce accurate answers efficiently. The computational methods that a student uses should be based on mathematical ideas that the student understands well, including the structure of the base-ten number system. (NCTM, 2000, p. 152)

Computation skills should be regarded as tools that further understanding, not as a substitute for understanding. (Paulos, 1991, p. 53)

When students merely memorize procedures, they may fail to understand the deeper ideas that could make it easier to remember—and apply—what they learn. Understanding makes it easier to learn skills, while learning procedures can strengthen and develop mathematical understanding. (NRC, 2002, p. 13)

Study results indicate that almost all children can and do invent strategies and that this process of invention (especially when it comes *before* learning standard algorithms) may have multiple advantages. (NCTM, 2002a, p. 93)

Teaching Implications

To support an initial understanding for students in regard to understanding multiple addition strategies, the following are ideas and questions to consider in conjunction with the research.

Focus Through Instruction

- Focusing on understanding multidigit addition methods results in much higher levels of correct use of methods.
- Students need visuals to understand the meanings of hundreds, tens, and ones. These meanings need to be related to the oral and written numerical methods developed in the classroom.
- Number lines and hundreds grids support counting-list methods the most effectively.
- Decomposition methods are facilitated by objects that allow children to physically add or remove different quantities (e.g., base-ten blocks).
- Students who believe there are several correct methods for adding numbers often show higher engagement levels.
- When children solve multidigit addition and subtraction problems, two types of problem-solving strategies are commonly used: invented strategies and standard algorithms. Although standard algorithms can simplify calculations, the procedures can be used without understanding, and multiple procedural issues can exist.
- Both invented and standard algorithms can be analyzed and compared, helping students understand the nature and properties of the operation, place-value concepts for numbers, and characteristics of efficient methods and strategies.

Questions to Consider (when working with students as they develop and/or interpret a variety of algorithms)

- When exploring and/or inventing algorithms, do students consider the generalizability of the method?
- Are students able to decompose and recompose the type of number they are operating with?
- Can students explain why a strategy results in the correct answer?
- When analyzing a strategy or learning a new method, do students focus on properties of numbers and the underlying mathematics rather than just memorizing a step-by-step procedure?
- Do students use a variety of estimation strategies to check the reasonableness of the results?

Teacher Sound Bite

I struggle to know what methods students bring with them each year when transitioning to my fourth-grade class. These strategy probes help me consider my students' level of comfort and experience with a variety of ways of adding numbers and whether they have more than just an understanding of the steps a procedure.

Curriculum Topic Study and Uncovering Student Thinking

What's Your Addition Strategy?

Keeley, P., & Rose, C. (2006). *Mathematics curriculum topic study: Bridging the gap between standards and practice.* Thousand Oaks, CA: Corwin. (Addition and Subtraction, p. 111).

Related Elementary Probes:

Rose, C., & Arline, C. (2009). *Uncovering student thinking in mathematics, grades 6–12: 30 formative assessment probes for the secondary classroom.* Thousand Oaks, CA: Corwin. (Variation: What's Your Addition Strategy? Decimals, p. 82; Fractions, p. 83).

Additional References for Research and Teaching Implications

McREL. (2002). *EDThoughts: What we know about mathematics teaching and learning.* Bloomington, IN: Solution Tree. (pp. 82–83).

National Council of Teachers of Mathematics. (2000). *Principles and standards for school mathematics.* Reston, VA: Author. (p. 152).

National Council of Teachers of Mathematics. (2002). *Lessons learned from research.* Reston, VA: Author. (pp. 93–100).

National Council of Teachers of Mathematics. (2003). *Research companion to principles and standards for school mathematics.* Reston, VA: Author. (pp. 68–84).

National Research Council. (2002). *Helping children learn mathematics.* Washington, DC: National Academy Press. (pp. 11–13).

National Research Council. (2005). *How students learn: Mathematics in the classroom.* Washington, DC: National Academy Press. (pp. 223–231).

Paulos, J. A. (1991). *Beyond numeracy.* New York: Vintage. (pp. 52–55).

Student Responses to What's Your Addition Strategy?

Sample Responses: Sam's Method

Student 1: "Sam is adding the easy way. First, add 6 and 4, put down the 0, and carry the 1, and then add 1 and 3 and 5, to get 9. Write 9 in front of the zero, so 90 is the answer."

Student 2: "Sam just has the answer, so I can't really tell how it got there. I also don't know why a little baby one is between the 3 and 4. Think it might be just a mistake."

Sample Responses: Julie's Method

Student 3: "First, Julie just added 30 plus 50 and wrote 80. Then, she added 6 and 4 and got 10. So, it's easy. 80 and 10 is 90."

Sample Responses: Pete's Method

Student 4: "My friend last year always used this way. I told him he was crazy, but it works for him. He likes rearranging numbers so the problem is different."

Sample Responses: Lisa's Method

Student 5: "I'm not really sure what Lisa is doing, but if you add up 20 + 10 + 4 in that row then you get 34. Just ignore the other row though because that one doesn't add up to any part of the problem."

PROBE 15A: VARIATION: WHAT'S YOUR ADDITION STRATEGY?

Sam, Julie, Pete, and Lisa each added the numbers 234 and 456.

1. *Circle* the method that most closely matches how you solved the problem.

2. *Explain* whether *each* method makes sense mathematically.

A) Sam's Method $\begin{array}{r} 2^13\,4 \\ +\ 4\,5\,6 \\ \hline 6\,9\,0 \end{array}$	
B) Julie's Method $\begin{array}{r} 2\,3\,4 \\ +\ 4\,5\,6 \\ \hline 6\,0\,0 \\ 8\,0 \\ \hline 1\,0 \\ \hline 6\,9\,0 \end{array}$	
C) Pete's Method $234 + 456$ $\quad +6 \quad -6$ $240 + 450$ $\quad\quad = 690$	
D) Lisa's Method $456 + 234$ $456 + 200 = 656$ $656 + 30 = 686$ $686 + 4 = 690$	

PROBE 15B: VARIATION: WHAT'S YOUR ADDITION STRATEGY?

Sam, Julie, Pete, and Lisa each added the numbers 1.6 and 0.8.

1. *Circle* the method that most closely matches how you solved the problem.

2. *Explain* whether *each* method makes sense mathematically.

A) Sam's Method $1.6 + .8$ $= 1 + .6 + .4 + .4$ $= 1 + 1 + .4$ $= 2.4$	
B) Julie's Method $\,^11.6$ $+\ \ 0.8$ $\overline{\ 2.4}$	
C) Pete's Method $1.6 + 0.8$ $+.4 - \ .4$ $2 + .4$ $= 2.4$	
D) Lisa's Method $= 1.6 + 0.8$ $= 1 + .6 + .8$ $= 1 + 1.4$ $= 2.4$	

PROBE 16: WHAT'S YOUR SUBTRACTION STRATEGY?

Sam, Julie, Pete, and Lisa each subtracted 28 from 67.

1. *Circle* the method that most closely matches how you solved the problem.

2. *Explain* whether *each* method makes sense mathematically.

A) Sam's Method $\quad 67$ $\underline{-28}$ $\underline{+40}$ $\quad\underline{-1}$ $\quad 39$	
B) Julie's Method $67 - 28$ $67 - 20 = 47$ $47 - 10 = 37$ $37 + 2 = 39$	
C) Pete's Method $67 - 28$ $+2 + 2$ $69 - 30$ $\quad = 39$	
D) Lisa's Method $\overset{5}{\cancel{6}}\overset{1}{7}$ $\underline{-28}$ $\quad 39$	

TEACHER NOTES: WHAT'S YOUR SUBTRACTION STRATEGY?

Questioning for Student Understanding

When subtracting, can students apply and understand a variety of different strategies?

K–2	3–5

Uncovering Understanding

What's Your Subtraction Strategy? Content Standard: Number and Operation

Important Note: Prior to giving students the probe, ask them to individually solve the indicated problem. If time allows, ask them to solve the problem in at least two or three different ways.

Variations/Adaptations:

o Subtraction Strategies: Three-Digit Numbers

o Subtraction Strategies: Decimals

Examining Student Work

Student answers may reveal *misunderstandings* regarding methods of subtracting, including a lack of *conceptual understanding* of the properties of numbers. Responses also may reveal a common misconception that there is only one correct algorithm for each operation or that, once comfortable with a method, there is no need to understand other methods.

- Sam's Method: This method of subtracting by the place and then adding the results is common in many elementary mathematics programs. Students who have only experienced a standard algorithm are typically confused because of the use of negative numbers. These students often refer to the incorrect idea that "you can never subtract a bigger number from a smaller number." (See Student Responses 1 and 2.)
- Julie's Method: This strategy consists of holding the minuend constant and breaking the subtrahend into place-value parts and subtracting on one part at a time. Students who have experience subtracting on an open number line are typically able to mathematically explain this method of subtraction. (See Student Response 3.)
- Pete's Method: This strategy is the least recognized by upper-elementary students in terms of generalizing a method of adding or subtracting like amounts from the numbers to keep the difference between the numbers

constant. The subtracting can be done mentally once the subtrahend is a multiple of 100. (See Student Response 4 and 5.)

- Lisa's Method: This method is usually recognized by upper-elementary students, although in some situations students may not have been introduced to the standard U.S. algorithm. Those who have no experience with the algorithm indicate a lack of understanding of the notation but can typically relate the method to regrouping with manipulatives. Those students who recognize the algorithm often do not demonstrate place-value understanding. (See Student Response 5.)

<u>S</u>eeking Links to Cognitive Research

Student errors when operating on whole numbers suggest students interpret and treat multi-digit numbers as single-digit numbers placed adjacent to each other, rather than using place-value meanings for digits in different positions. (AAAS, 1993, p. 358)

Students can use roughly three classes of effective methods for multi-digit addition and subtraction, although some methods are mixtures. *Counting-list methods* are extensions of the single-digit counting methods. Children initially may count large numbers by ones, but these unitary methods are highly inaccurate and are not effective. All children need to be helped as rapidly as possible to develop prerequisites for methods using tens. These methods generalize readily to counting on or up by hundreds but become unwieldy for larger numbers. In *decomposing methods,* children decompose numbers so that they can add or subtract the like units (e.g., add tens to tens, ones to ones, hundreds to hundreds, etc.). These methods generalize easily to very large numbers. *Recomposing methods* are like the make-a-ten or doubles methods. The solver changes both numbers by giving some amount of one number to another number (i.e., in adding) or by changing both numbers equivalently to maintain the same difference (i.e., in subtracting). (NCTM, 2003, p. 79)

When students merely memorize procedures, the may fail to understand the deeper ideas. When subtracting, for example, many children subtract the smaller number from the larger in each column, no matter where it is. (NRC, 2002, p. 13)

By the end of the 3–5 grade band, students should be computing fluently with whole numbers. Computational fluency refers to having efficient and accurate methods for computing. Students exhibit computational fluency when they demonstrate flexibility in the computational methods they choose, understand and can explain these methods, and produce accurate answers efficiently. The computational methods that a student uses should be based on mathematical ideas that the student understands well, including the structure of the base-ten number system. (NCTM, 2000, p. 152)

When students merely memorize procedures, they may fail to understand the deeper ideas that could make it easier to remember—and apply—what they learn. Understanding makes it easier to learn skills, while learning procedures can strengthen and develop mathematical understanding. (NRC, 2002, p. 13)

Study results indicate that almost all children can and do invent strategies and that this process of invention (especially when it comes *before* learning standard algorithms) may have multiple advantages. (NCTM, 2002a, p. 93)

*T*eaching Implications

To support a deeper understanding for students in secondary grades in regard to number and operations, the following are ideas and questions to consider in conjunction with the research.

Focus Through Instruction

- Focusing on understanding multidigit subtraction methods results in much higher levels of correct use of methods.
- Students need visuals to understand the meanings of hundreds, tens, and ones. These meanings need to be related to the oral and written numerical methods developed in the classroom.
- Number lines and hundreds grids support counting-list methods the most effectively.
- Decomposition methods are facilitated by objects that allow children to physically add or remove different quantities (e.g., base-10 blocks).
- Student's who believe there are several correct methods for subtracting numbers often show higher engagement levels.
- When children solve multidigit subtraction problems, two types of problem-solving strategies are commonly used: invented strategies and standard algorithms. Although standard algorithms can simplify calculations, the procedures can be used without understanding, and multiple procedural issues can exist.
- Both invented and standard algorithms can be analyzed and compared, helping students understand the nature and properties of the operation, place-value concepts for numbers, and characteristics of efficient methods and strategies.

Questions to Consider (when working with students as they develop or interpret a variety of algorithms)

- Do students understand the different meanings of subtraction (e.g., take away, difference, distance)?
- When exploring or inventing algorithms, do students consider the generalizability of the method?

- Are students able to decompose and recompose the type of number they are operating with?
- Can students explain why the strategy results in the correct answer?
- When analyzing a strategy or learning a new method, do students focus on properties of numbers and the underlying mathematics rather than just memorizing a step-by-step procedure?
- Do students use a variety of estimation strategies to check the reasonableness of the results?

Teacher Sound Bite

It seems that every year students have difficulty with subtracting numbers using multiple methods even though for addition the multiple methods do not seem to be a problem. Once I considered the idea that subtraction has multiple meanings, I was much better able to show visuals and use manipulatives to model the operation in different ways. It pleased me that this year my students added many more visual representations to their postassessment responses (gave probe initially, spent warm-ups on the concept for five days, and redistributed the probe as a postassessment to allow students to make changes).

Curriculum Topic Study and Uncovering Student Thinking

What's Your Subtraction Strategy?

Keeley, P., & Rose, C. (2006). *Mathematics curriculum topic study: Bridging the gap between standards and practice.* Thousand Oaks, CA: Corwin. (Addition and Subtraction, p. 111).

Elementary Probes:

Rose, C., & Arline, C. (2009). *Uncovering student thinking in mathematics, grades 6–12: 30 formative assessment probes for the secondary classroom.* Thousand Oaks, CA: Corwin. (Variation: What's Your Subtraction Strategy? Decimals, p. 90; Fractions, p. 91).

Additional References for Research and Teaching Implications

McREL. (2002). *EDThoughts: What we know about mathematics teaching and learning.* Bloomington, IN: Solution Tree. (pp. 82–83).

National Council of Teachers of Mathematics. (2002). *Lessons learned from research.* Reston, VA: Author. (pp. 93–100).

National Council of Teachers of Mathematics. (2003). *Research companion to principles and standards for school mathematics.* Reston, VA: Author. (pp. 68–84).

National Research Council. (2002). *Helping children learn mathematics.* Washington, DC: National Academy Press. (pp. 11–13).

National Research Council. (2005). *How students learn: Mathematics in the classroom.* Washington, DC: National Academy Press. (pp. 223–231).

Student Responses to What's Your Subtraction Strategy?

Sample Responses: Sam's Method

Student 1: "At our school, we name this the partial difference method, not Sam's method. I explain it like this. First, take 60 – 20, and write down 40. Second, look at the ones numbers. If the bottom is smaller than, write down + the number, but if it is bigger than, write down – the number. Do what it says to do, and you're done."

Student 2: "I don't know how 67 – 28 + 40 – 1 comes out to 39. I think it comes out to 88."

Sample Responses: Julie's Method

Student 3: "Julie likes benchmark numbers since she's only subtracting by 20 then 10, but then that's too much, so the 2 is added back in."

Sample Responses: Pete's Method

Student 4: "Haven't seen this, so not sure. 67 + 2 is 69, and 28 + 2 is 30, so the math's right anyway. Not sure about why the 2 and not some other number."

Sample Responses: Lisa's Method

Student 5: "I use this one all the time. It's the easiest. Cross out the 6 and make it a 5, so that 7 is now a 17. Now you can do the work."

Student 6: "The 51 and 67 together confuses me, but Lisa must make sense of it. Her answer's right."

PROBE 16A: VARIATION: WHAT'S YOUR SUBTRACTION STRATEGY?

Sam, Julie, Pete, and Lisa each subtracted 284 from 672.

1. *Circle* the method that most closely matches how you solved the problem.

2. *Explain* whether *each* method makes sense mathematically.

A) Sam's Method 672 − 284 + 400 − 10 − 2 388	
B) Julie's Method 672 − 284 672 − 200 = 472 472 − 80 = 392 392 − 4 = 388	
C) Pete's Method 672 − 284 + 6 + 6 678 − 290 + 10 + 10 688 − 300 = 388	
D) Lisa's Method 5 6 16 7 12 − 284 3 8 8	

PROBE 16B: VARIATION: WHAT'S YOUR SUBTRACTION STRATEGY?

Sam, Julie, Pete, and Lisa each subtracted 2.8 from 6.7.

1. *Circle* the method that most closely matches how you solved the problem.

2. *Explain* whether *each* method makes sense mathematically.

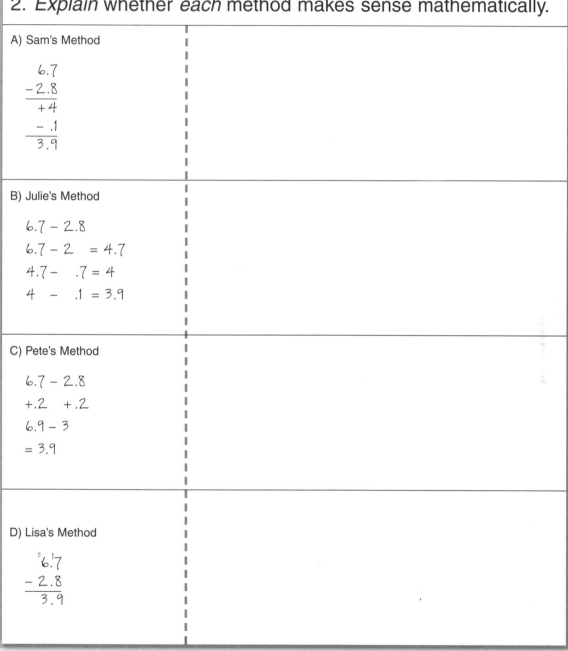

A) Sam's Method

$$\begin{array}{r} 6.7 \\ -2.8 \\ \hline +4 \\ -.1 \\ \hline 3.9 \end{array}$$

B) Julie's Method

$6.7 - 2.8$
$6.7 - 2 \ \ = 4.7$
$4.7 - \ \ .7 = 4$
$4 \ \ - \ \ .1 = 3.9$

C) Pete's Method

$6.7 - 2.8$
$+.2 \ \ +.2$
$6.9 - 3$
$= 3.9$

D) Lisa's Method

$$\begin{array}{r} {}^{5}6.{}^{1}7 \\ -2.8 \\ \hline 3.9 \end{array}$$

PROBE 17: WHAT'S YOUR MULTIPLICATION STRATEGY?

Sam, Julie, Pete, and Lisa each multiplied 28 by 17.

1. *Circle* the method that most closely matches how you solved the problem.

2. *Explain* whether *each* method makes sense mathematically.

A) Sam's Method	
$\begin{array}{r} {}^{5}28 \\ 17 \\ \hline {}^{1}196 \\ +\ 280 \\ \hline 476 \end{array}$	

B) Julie's Method	
(lattice multiplication grid)	

C) Pete's Method	
$28 \times 10 = 280$ $28 \times 5 = 140$ $28 \times 2 = 56$ $280 + 140 + 56 = 476$	

D) Lisa's Method	
(area model: 20 and 8 across; 10 and 7 down; 200, 80, 140, 56) $200 + 80 + 140 + 56 = 476$	

TEACHER NOTES: WHAT'S YOUR MULTIPLICATION STRATEGY?

Questioning for Student Understanding

When multiplying, can students apply and understand a variety of different strategies?

K–2	3–5

Uncovering Understanding

What's Your Multiplication Strategy? Content Standard: Number and Operation

Important Note: Prior to giving students the probe, ask them to individually solve the indicated problem. If time allows, ask them to solve the problem in at least two or three different ways.

Examining Student Work

Student answers may reveal *misunderstandings* regarding methods of multiplication including a lack of *conceptual understanding* of the properties of numbers. Responses also may reveal a common misconception that there is only one correct algorithm for each operation or that, once comfortable with a method, there is no need to understand other methods.

- Sam's Method: As the typical U.S. standard algorithm, this method is labeled as making sense by most students. Although familiar with the method, most students give a procedural explanation of the steps using incorrect place-value language. Those students who have not been introduced to the algorithm have great difficulty trying to interpret the steps. (See Student Response 1.)
- Julie's Method: The lattice method, which has been around for centuries, is a process taught in some of the K–5 elementary mathematics programs. In the chart, partial products are laid out in place-value diagonals. Students who successfully use this method may lack concepts of place value, seeing the numbers as single digits. (See Student Response 2.)
- Pete's Method: The method of keeping one factor whole while breaking up the other into several addends relies on known facts, such as

working with multiples of 10, and doubling and halving. (See Student Responses 3.)

- Lisa's Method: This method breaks the numbers into place-value parts, resulting in partial products that are then added together. Students who understand this method are often better able to apply this knowledge to the multiplication of polynomials. (See Student Responses 4.)

Seeking Links to Cognitive Research

Student errors suggest students interpret and treat multi-digit numbers as single-digit numbers placed adjacent to each other, rather than using place-value meanings for the digits in different positions. (AAAS, 1993, p. 358)

Researchers have reported a preliminary learning progression of multi-digit methods in which teachers fostered students' invention of algorithms. These methods moved from (a) direct modeling with objects or drawings, to (b) written methods involving repeatedly adding and/or doubling, to (c) partitioning methods. (NCTM, 2003, p. 84)

The multiplication algorithm currently most prevalent is a complex method that is not easy to understand or to carry out. It demands high levels of skill in multiplying a multi-digit number by a single-digit number within an embedded format in which multiplying and adding alternate. The meaning and scaffolding of sub-steps [have] been sacrificed, using aligning methods that keep the steps organized by correct place value without requiring any understanding of what is actually happening with the ones, tens, and hundreds. (NCTM, 2003, p. 85)

Researchers and experienced teachers alike have found that when students are encouraged to develop, record, explain, and critique one another's strategies for solving computational problems, a number of important kinds of learning can occur. For students to become fluent in arithmetic computation, they must have efficient and accurate methods that are supported by an understanding of numbers and operations. (NCTM, 2000, p. 35)

Teaching Implications

To support a deeper understanding for students in secondary grades in regard to number and operations, the following are ideas and questions to consider in conjunction with the research.

Focus Through Instruction

- Some of the practice needed to master multidigit multiplication skills can be carried out using context-free numbers, but understanding should be based primarily through solving problems in which the answers matter within a given situation.
- Memorization of algorithms by drill does not lend easily to understanding how the numbers within the problem provide clues as to the appropriate choice of a particular method.
- Students who lack ability to state multiplication facts should still have experience with modeling multiple multidigit multiplication methods.
- Regardless of the particular method used, students should be able to explain their method, understand that many methods exist, and see the usefulness of methods that are efficient and accurate.

Questions to Consider
(when working with students as they
develop or interpret a variety of algorithms)

- Do student understand the multiple meanings and representations of multiplication?
- When exploring or inventing algorithms, do students consider the generalizability of the method?
- Are students able to decompose and recompose the type of number they are operating with?
- Can students explain why the strategy results in the correct answer?
- When analyzing a strategy or learning a new method, do students focus on properties of numbers and the underlying mathematics rather than just memorizing a step-by-step procedure?
- Do students use a variety of estimation strategies to check the reasonableness of the results?

Teacher Sound Bite

I often tell my students over and over to find one strategy they like and then to stick with it. I am beginning to wonder though if this limits their ability to estimate and to want to consider other strategy choices. By using this probe and developing similar sets of items to give every month or so, I now plan to keep the discussion about multiple strategies *alive* in a way that won't take too much time.

Curriculum Topic Study and Uncovering Student Thinking

What's Your Multiplication Strategy?

Keeley, P., & Rose, C. (2006). *Mathematics curriculum topic study: Bridging the gap between standards and practice.* Thousand Oaks, CA: Corwin. (Multiplication and Division, p. 125).

Related Elementary Probe:

Rose, C., & Arline, C. (2009). *Uncovering student thinking in mathematics, grades 6–12: 30 formative assessment probes for the secondary classroom.* Thousand Oaks, CA: Corwin. (Variation: What's Your Decimal Multiplication Strategy? p. 98).

Additional References for Research and Teaching Implications

American Association for the Advancement of Science. (1993). *Benchmarks for science literacy.* New York: Oxford University Press. (pp. 350, 358).

Bay Area Mathematics Task Force. (1999). *A mathematics sourcebook for elementary and middle school teachers.* Novato, CA: Arena Press. (pp. 71–79).

National Council of Teachers of Mathematics. (1993). *Research ideas for the classroom: Middle grades mathematics.* New York: Macmillan. (pp. 99–115, 137–158).

National Council of Teachers of Mathematics. (2000). *Principles and standards for school mathematics.* Reston, VA: Author. (pp. 32–36).

National Council of Teachers of Mathematics. (2003). *Research companion to principles and standards for school mathematics.* Reston, VA: Author. (pp. 84–91, 114–120).

National Research Council. (2001). *Adding it up: Helping children learn mathematics.* Washington, DC: National Academy Press. (pp. 231–241).

National Research Council. (2002). *Helping children learn mathematics.* Washington, DC: National Academy Press. (pp. 11–13).

National Research Council. (2005). *How students learn: Mathematics in the classroom.* Washington, DC: National Academy Press. (pp. 309–349).

Student Responses to What's Your Multiplication Strategy?

Sample Responses: Sam's Method

Student 1: "Sam's is the normal method, so it makes sense."

Sample Responses: Julie's Method

Student 2: "Yes, the multiplication is in the right order, and it is added up right."

Sample Responses: Pete's Method

Student 3: "No, I never learned this method or maybe never followed it. This is too confusing."

Sample Responses: Lisa's Method

Student 4: "This makes sense because it's so simple to multiple and then add."

PROBE 17A: VARIATION: WHAT'S YOUR MULTIPLICATION STRATEGY?

Sam, Julie, Pete, and Lisa each multiplied 12 by 8.

1. *Circle* the method that most closely matches how you solved the problem.

2. *Explain* whether *each* method makes sense mathematically.

A) Sam's Method	
$\begin{array}{r} {}^{1}12 \\ \times\ 8 \\ \hline 96 \end{array}$	

B) Julie's Method	
(area/lattice diagram)	

C) Pete's Method	
12×8 $10 \times 8 = 80$ $2 \times 8 = 16$ $80 + 16 = 96$	

D) Lisa's Method	
(area model: 10 and 2 across; 4 and 4 down; cells 40, 8, 40, 8) $40 + 40 + 8 + 8 = 96$	

PROBE 18: WHAT'S YOUR DIVISION STRATEGY?

Sam, Julie, Pete, and Lisa each divided 84 by 7.

1. *Circle* the method that most closely matches how you solved the problem.

2. *Explain* whether *each* method makes sense mathematically.

A). Sam's Method

$84 - 14 = 70$
$70 - 14 = 56$
$56 - 14 = 42$
$42 - 14 = 28$
$28 - 14 = 14$
$14 - 14 = 0$
$6 \times 2 = 12$

B) Julie's Method

```
    7) 84
   -   70     10
 2    14
   -   14
     0
 10 + 2 = 12
```

C) Pete's Method

D) Lisa's Method

```
      12
  7) 84
      7
     14
     14
      0
```

TEACHER NOTES: WHAT'S YOUR DIVISION STRATEGY?

Questioning for Student Understanding

When dividing, can students apply and understand a variety of different strategies?

K–2	3–5

Uncovering Understanding

What's Your Division Strategy? Content Standard: Number and Operations

Important Note: Prior to giving students the probe, ask them to individually solve the indicated problem. If time allows, ask them to solve the problem in at least two or three different ways.

Examining Student Work

Student answers may reveal *misunderstandings* regarding methods of multiplication including a lack of *conceptual understanding* of the properties of numbers. Responses also may reveal a common misconception that there is only one correct algorithm for each operation or that, once comfortable with a method, there is no need to understand other methods.

- Sam's Method: This method involves repeatedly subtracting 14 from the number. Because 14 is double the dividend of 7, the student doubles the number of 14s that were subtracted. (See Student Responses 1 and 2.)
- Julie's Method: Often referred to as partial quotients, this method relies on repeated subtraction of known parts. (See Student Responses 3 and 4.)
- Pete's Method: This equal-share method indicates a conceptual understanding of division but can easily become cumbersome with larger numbers. (See Student Responses 5 and 6.)
- Lisa's Method: As the typical U.S. standard algorithm, this method is labeled as making sense by most students. Although familiar with the method, most students give a procedural explanation of the steps using incorrect place-value language. Those students who have not been introduced to the algorithm have great difficulty trying to interpret the steps. (See Student Responses 7 and 8.)

Seeking Links to Cognitive Research

The usual U.S. division algorithm has two aspects that create difficulties for students. First, it requires them to determine exactly the maximum copies of the divisor that they can take from the dividend. Second, it creates no sense of the size of the answers that students are writing; in fact, they are always multiplying by single digits. Thus, students have difficulty gaining experience with estimating the correct order of magnitude of answers in division. (NCTM, 2003, pp. 85–86)

Student errors suggest students interpret and treat multi-digit numbers as single-digit numbers placed adjacent to each other, rather than using place-value meanings for the digits in different positions. (AAAS, 1993, p. 358)

An example of disconnection in proficiency is students' tendency to compute with written symbols in a mechanical way without considering what the symbols mean. A survey of students' performance showed that the most common error for the addition problem $4 + .3 = ?$ is $.7$, which is given by 68% of sixth graders and 51% of fifth and seventh graders (Hiebert & Wearne, 1986). The errors show that many students have learned rules for manipulating symbols without understanding what those symbols mean or why the rules work. (NRC, 2001, p. 234)

Researchers have reported a preliminary learning progression of multidigit methods in which teachers fostered students' invention of algorithms. These methods moved from (a) direct modeling with objects or drawings, to (b) written methods involving repeatedly adding and/or doubling, to (c) partitioning methods. (NCTM, 2003, p. 84)

Researchers and experienced teachers alike have found that when students are encouraged to develop, record, explain, and critique one another's strategies for solving computational problems, a number of important kinds of learning can occur. For students to become fluent in arithmetic computation, they must have efficient and accurate methods that are supported by an understanding of numbers and operations. (NCTM, 2000, p. 35)

Teaching Implications

To support a deeper understanding for students in secondary grades in regard to number and operations, the following are ideas and questions to consider in conjunction with the research.

Focus Through Instruction

- Some of the practice needed to master division skills can be carried out using context-free numbers, but understanding should be based primarily through solving problems in which the answers matter within a given situation.
- Discuss how the numbers in the problem provide clues as to the appropriate choice of a particular method.
- Students who lack ability to state multiplication facts should still have experience with modeling multiple multidigit division methods.
- Regardless of the particular method used, students should be able to explain their method, understand that many methods exist, and see the usefulness of methods that are efficient and accurate.

Questions to Consider (when working with students as they develop or interpret a variety of algorithms)

- Do student understand the multiple meanings and representations of division?
- When exploring or inventing algorithms, do students consider the generalizability of the method?
- Are students able to decompose and recompose the type of number they are operating with?
- Can students explain why the strategy results in the correct answer?
- When analyzing a strategy or learning a new method, do students focus on properties of numbers and the underlying mathematics rather than just memorizing a step-by-step procedure?
- Do students use a variety of estimation strategies to check the reasonableness of the results?

Teacher Sound Bite

As a Gifted and Talented teacher, I find many of the students prefer to learn the standard algorithm and often do not care to understand why the algorithm always works. I find the strategy probes useful as way to encourage them to compare and contrast strategies they otherwise wouldn't see or analyze closely. I also extend the probe by asking the students to create another algorithm not already shown. This pushes those students who can quickly make sense of work shown.

Additional References for Research and Teaching Implications

American Association for the Advancement of Science. (1993). *Benchmarks for science literacy.* New York: Oxford University Press. (pp. 350, 358).

Bay Area Mathematics Task Force. (1999). *A mathematics sourcebook for elementary and middle school teachers.* Novato, CA: Arena Press. (pp. 71–79).

National Council of Teachers of Mathematics. (1993). *Research ideas for the classroom: Middle grades mathematics.* New York: Macmillan. (pp. 99–115, 137–158).

National Council of Teachers of Mathematics. (2000). *Principles and standards for school mathematics.* Reston, VA: Author. (pp. 32–36).

National Council of Teachers of Mathematics. (2003). *Research companion to principles and standards for school mathematics.* Reston, VA: Author. (pp. 84–91, 114–120).

National Research Council. (2001). *Adding it up: Helping children learn mathematics.* Washington, DC: National Academy Press. (pp. 231–241).

Curriculum Topic Study and Uncovering Student Thinking

What's Your Division Strategy?

Keeley, P., & Rose, C. (2006). *Mathematics curriculum topic study: Bridging the gap between standards and practice.* Thousand Oaks, CA: Corwin. (Multiplication and Division, p. 125).

Related Elementary Probe:

Rose, C., & Arline, C. (2009). *Uncovering student thinking in mathematics, grades 6–12: 30 formative assessment probes for the secondary classroom.* Thousand Oaks, CA: Corwin. (Variation: What's Your Decimal Division Strategy? p. 105).

National Research Council. (2002). *Helping children learn mathematics.* Washington, DC: National Academy Press. (pp. 11–13).

National Research Council. (2005). *How students learn: Mathematics in the classroom.* Washington, DC: National Academy Press. (pp. 309–349).

Student Responses to What's Your Division Strategy?

Sample Responses: Sam's Method

Student 1: "I think it is broken down into simple terms by subtracting 14s until he hit 0. He is dividing by 7, not 14, so if there are six 14s, there would have to be twelve 7s."

Student 2: "Sam doesn't make sense because there aren't any 7s in his work."

Sample Responses: Julie's Method

Student 3: "I learned this last year as the magic 7 way. Subtract what you know first, and then it is an easier number to think about."

Student 4: "I did it like Lisa, so Julie's doesn't make sense, but maybe she broke it apart?"

Sample Responses: Pete's Method

Student 5: "This does make sense because it is the simplest way to divide a number by having objects to spread out, so it's like all 7 people get the same amount."

Student 6: "My head hurts looking at it. Too many circles and lines."

Sample Responses: Lisa's Method

Student 7: "This is what we always do. 7 goes into 8 one time, so put a 1 above the 8. 8 – 7 is 1 left over, so bring down the 4 to make 14. 7 goes into 14 two times, so put the 2 above where the 4 is. Since 14 – 14 is 0, it's done."

Student 8: "I haven't learned this yet, but I know some people in my class do know it."

PROBE 19: IS IT AN ESTIMATE?

Decide whether each student's solution is a correct way to estimate the answer to 46 + 24 + 33.

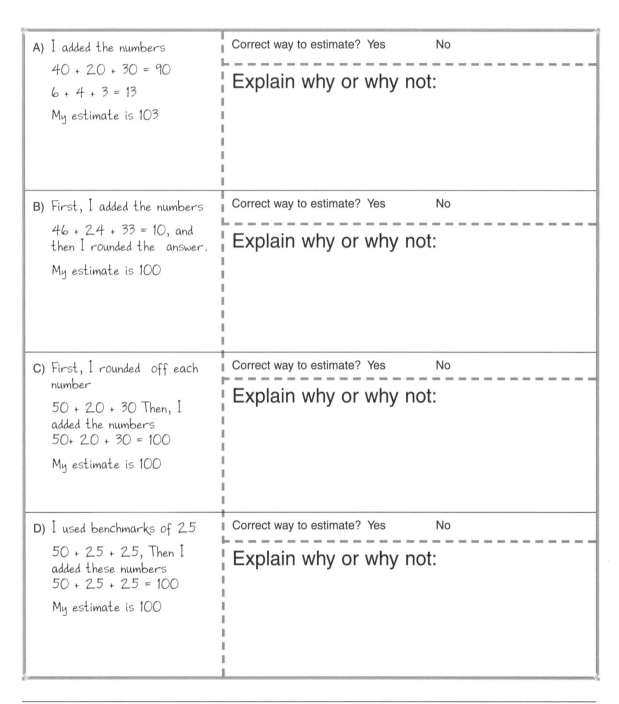

A) I added the numbers

40 + 20 + 30 = 90

6 + 4 + 3 = 13

My estimate is 103

Correct way to estimate? Yes   No

Explain why or why not:

B) First, I added the numbers

46 + 24 + 33 = 10, and then I rounded the answer.

My estimate is 100

Correct way to estimate? Yes   No

Explain why or why not:

C) First, I rounded off each number

50 + 20 + 30 Then, I added the numbers
50 + 20 + 30 = 100

My estimate is 100

Correct way to estimate? Yes   No

Explain why or why not:

D) I used benchmarks of 25

50 + 25 + 25, Then I added these numbers
50 + 25 + 25 = 100

My estimate is 100

Correct way to estimate? Yes   No

Explain why or why not:

TEACHER NOTES: IS IT AN ESTIMATE?

Questioning for Student Understanding

Do students understand there are multiple methods of estimating the sum of three 2-digit numbers?

K–2	3–5

Uncovering Understanding

Is It an Estimate? Content Standard: Number and Operations

Note: Prior to giving students the probe, ask them to individually solve the indicated problem (*Estimate* the sum of 46 + 24 + 33).

Examining Student Work

Student answers may reveal lack of *conceptual understanding* regarding methods of estimation. Responses may reveal a *common misunderstanding* that there is only one correct way to estimate and *common procedural errors*, such as that estimation is rounding an answer to a computation problem.

- Method 1: Students who choose this method view estimation as solving a problem or solving a problem in an alternative way. (See Student Responses 1.)
- Method 2: Students who choose this method view estimation and rounding as the same procedure. (See Student Responses 2.)
- Method 3: Student who choose this method are demonstrating an understanding of estimation but may not realize that how precise an estimate is needed depends on the context of the situation. This partial understanding can be indicated by responses such as "This is an estimate but not as good as rounding to the nearest 10." (See Student Responses 3.)
- Method 4: Students who choose this method along with Method 3 are demonstrating an understanding of estimation. Those who do not are often associating estimation only with rounding to the nearest 10 or 100. (See Student Responses 4.)

Seeking Links to Cognitive Research

In Grades 3–5, all students should develop and use strategies to estimate the results of whole-number computations and to judge the reasonableness of such results. (NCTM, 2000, p. 148)

Students in grades 3–5 will need to be encouraged to routinely reflect on the size of an anticipated solution. Instructional attention and frequent modeling by the teacher can help students develop a range of computational estimation strategies including flexible rounding, the use of benchmarks, and front-end strategies. (NCTM, 2000, p. 155)

Third graders can seem tied to the mechanical methods of rounding to the nearest 10 or 100 and are not able to accept different close or rounded numbers for different situations. (NCTM, 2003, p. 145)

Work with strategies such as using only the *front-end* digits, *rounding,* or the *use of special numbers* can begin in 3rd or 4th grade. The front end strategy is a good place to start with younger students because it clearly illustrates the approximateness of estimation. (NCTM, 2003a, p. 145)

Students often think it is inferior to exact computation and equate it with guessing (Sowder, 1992b), so that they do not believe estimation is useful (Sowder & Wheeler, 1989). (AAAS, 1993, p. 350)

Researchers attempting to identify and characterize the computational processes used by good estimators found three key processes. The first was called reformulation and referred to changing the numbers to other numbers that were easier to manage mentally. A second process was called translation and referred to changing the structure of the problem so that the operations could be more easily carried out mentally. The third process used was compensation where adjustments were made both during and after estimating. (NCTM, 1993b, pp. 43–44)

Teaching Implications

To support a deeper understanding for students in regard to numbers and operations, the following are ideas and questions to consider in conjunction with the research.

Focus Through Instruction

- Estimation skills can be learned, but only if teachers make sure that students have lots of practice. This happens if estimation is routinely treated as a standard part of problem solving.
- When students are frequently called upon to explain how they intend to calculate an answer before carrying it out, they view estimation as a natural part of the computation process.
- Instructional attention and frequent modeling by the teacher can help students develop a range of computational estimation strategies, including flexible rounding, the use of benchmarks, and front-end strategies.

- As with exact computation, sharing estimation strategies allows students access to others' thinking and provides many opportunities for rich class discussions.
- For students to become really skilled at estimation, it has to be incorporated into their regular instruction over several years.

Questions to Consider (when working with students as they grapple with the process of estimation)

- Are students using various methods of estimating to check reasonableness of computation results?
- Do students view estimation as more than just guessing?
- Do students understand when to overestimate versus underestimate?
- Do students have number sense regarding types of numbers they are estimating with?

Teacher Sound Bite

Based on prior experiences, some students think that because there are multiple strategies to estimate, any and all strategies that are labeled estimates have to be correct (maybe this relates to the research that indicates students often see estimation as guessing); and other students think that there is only one correct way to estimate, which is to round to the nearest multiple of 10. Using the probe provides a way for me to see how many students fall into these two common categories. It is then my job to constantly ask students to estimate and which type of estimate is required given the setting of the problem. By consistently asking students to estimate prior to calculating, I am modeling estimation as a method of checking an answer, but I also realized I needed to provide problems in which the estimate *is* the final answer.

Curriculum Topic Study and Uncovering Student Thinking

Is It an Estimate?

Keeley, P., & Rose, C. (2006). *Mathematics curriculum topic study: Bridging the gap between standards and practice.* Thousand Oaks, CA: Corwin. (Estimation, p. 195).

Additional References for Research and Teaching Implications

American Association for the Advancement of Science. (1989). *Science for all Americans: A Project 2061 report on literacy goals in science, mathematics, and technology.* Washington, DC: Author. (pp. 190–191).

American Association for the Advancement of Science. (1993). *Benchmarks for science literacy.* New York: Oxford University Press. (pp. 288–299, 350).

National Council of Teachers of Mathematics. (2000). *Principles and standards for school mathematics.* Reston, VA: Author. (pp. 32–36).

Student Responses to Is It an Estimate?

Sample Responses: Method 1

Student 1: "No: The estimate is not a benchmark number."

Sample Responses: Method 2

Student 2: "Yes: This is right because an estimate is a benchmark number."

Sample Responses: Method 3

Student 3: "Yes: Of all the ways, this is like how I do it. Use numbers close to 10s, then add."

Sample Responses: Method 4

Student 4: "No: You can't use those numbers. They should end in a zero."

PROBE 20: WHAT IS YOUR ESTIMATE?

Circle the best benchmark estimate.	Explain your thinking.
A) $$\frac{12}{13} + \frac{7}{8}$$ a) $\frac{1}{4}$ b) $\frac{1}{2}$ c) 1 d) 2 e) 19 f) 21 g) 40	
B) $$\frac{1}{8} + \frac{1}{10}$$ a) $\frac{1}{4}$ b) $\frac{1}{2}$ c) $\frac{3}{4}$ d) 1 e) 2 f) 18 g) 10	
C) $$\frac{4}{5} - \frac{1}{2}$$ a) $\frac{1}{4}$ b) $\frac{1}{2}$ c) $\frac{3}{4}$ d) −1 e) 0 f) 1 g) 3	

TEACHER NOTES: WHAT IS YOUR ESTIMATE?

<u>Q</u>uestioning for Student Understanding

Can students use estimation to choose the closest benchmark to addition and subtraction problems involving fractions?

K–2	3–5

<u>U</u>ncovering Understanding

What Is Your Estimate? Content Standard: Number and Operations

<u>E</u>xamining Student Work

The distracters may reveal *common misunderstandings* regarding large numbers, such as a lack of *conceptual understanding* of the size of a fraction and common errors related to inaccurate use of procedures.

- The correct answers are A, d; B, a; and C, b: Students who choose the correct answers are able to use benchmark numbers to provide a more-accurate estimate. (See Student Response 1.)
- Distracters for item A: Students who choose a, $\frac{1}{4}$, or b, $\frac{1}{2}$, typically believe that adding two fractions results in another fraction that is less than one. Students who choose c,1, typically believe that to add two fractions you would add the numerators, then the denominators, and use the result of $\frac{19}{21}$ to determine their choice of 1 as an estimate. Or, they believe that adding any two fractions should give an answer close to one whole. Students who choose e, 19, f, 21, or g, 40, are typically either adding the numerators, adding the denominators, or adding all of the numbers. (See Student Responses 2 and 3.)
- Distracters for item B: Students who choose b, $\frac{1}{4}$, or c, $\frac{1}{2}$, typically believe that adding two fractions results in another fraction that is less than one. Students who choose d,1 typically believe that adding any two fractions should be close to one whole. Students who choose e, 2, f, 18, or g, 20, are typically either adding the numerators, adding the denominators, or adding all of the numbers. (See Student Responses 4 and 5.)
- Distracters for item C: Students who choose a, $\frac{1}{4}$, typically believe that subtracting two fractions results in an answer close to zero, so they choose the smallest possible answer that is larger than zero. Students who choose c, $\frac{3}{4}$, typically choose the closest fraction to one whole. Students who choose d, −1, typically think subtracting fractions will result in a number less than zero. Students who choose e, 0, believe subtracting any two fractions results in an answer close to zero, or they are subtracting all of the numbers. Students who choose g, 3, are typically either subtracting the numerators or denominators. (See Student Responses 6 and 7.)

Seeking Links to Cognitive Research

Making judgments about answers is as much a part of computation as the calculation itself. Students need to develop estimation skills and the habit of checking answers against reality. (AAAS, 1993, p. 288)

In Grades 3 through 5, students can learn to compare fractions to familiar benchmarks such as $\frac{1}{2}$. And, as their number sense develops, students should be able to reason about numbers by, for instance, explaining that $\frac{1}{2} + \frac{3}{8}$ must be less than 1 because each addend is less than or equal to $\frac{1}{2}$. (NCTM, 2000, p. 33)

Elementary students make several errors when they operate on decimals and fractions (Benander & Clement, 1985; Kouba et al., 1988; Peck & Jencks, 1981; Wearne & Hiebert, 1988). These errors are due in part to the fact that students lack essential concepts about decimals and fractions and have memorized procedures that they apply incorrectly. (AAAS, 1993, pp. 358–359)

Students' difficulties indicate they do not perceive a fraction as a single quantity. Instead, they treat the numerator and denominator separately as a pair of whole numbers. (AAAS, 1993, p. 359)

Estimation of fraction computations is mostly tied to the concept of the operation and the concept of fraction. An algorithm is not required for making such estimates. (Van de Walle, 2007, p. 316)

Teaching Implications

To support a deeper understanding for students in secondary grades in regard to number and operations, the following are ideas and questions to consider in conjunction with the research.

Focus Through Instruction

- Visual images of fractions as fraction strips should help many students think flexibly in comparing fractions. Students may also be helped by thinking about the relative locations of fractions on a number line.
- Teachers can help students add and subtract fractions correctly by helping them develop meanings for *numerator* and *denominator* and *equivalence*, and by encouraging them to use benchmarks and estimation.
- In the lower grades, students should have had experiences in comparing fractions between zero and one in relation to such benchmarks as 0, $\frac{1}{4}$, $\frac{1}{2}$, $\frac{3}{4}$, and 1. In the middle grades, students should extend this experience to tasks in which they order and compare fractions, which many students find difficult.
- Repeated experience with computations in meaningful contexts will foster the skill of judging the appropriateness of an answer.
- Make sure that students have lots of practice estimating—this happens if estimation is routinely treated as a standard part of problem solving.

- When students are frequently called upon to explain how they intend to calculate an answer before carrying it out, they find that making step-by-step estimations is viewed as being difficult.
- Students should be encouraged to pinpoint a range of benchmark estimates (e.g., "I know it is smaller than ___ but larger than ___").

Questions to Consider (when working with students as they grapple with concepts related to fractions)

- Can students accurately judge the size of fractions in relation to common benchmarks, such as $\frac{1}{3}, \frac{1}{4}, \frac{1}{2}, \frac{3}{4}$, and so on?
- Can students use various models to represent fractions, including fraction strips, area models, and sets of objects?
- Can students explain how the size of the numerator and denominator of a fraction helps determine its size?
- Do students understand they do not need common denominators when estimating?

Teacher Sound Bite

Our district has recently put a lot of emphasis on use of benchmarks and estimates. The conversations have helped our transition group [a group of fourth- to seventh-grade teachers working on elementary-to-middle school transition to new math program] to realize that students' later difficulty with fraction operations related directly to students' inability to judge whether an end calculation was in the right ballpark. It's pretty hard to know if you have an answer that makes sense if you have no understanding of the meaning of the answer. This probe helped our transition group look across multiple grade levels to see if our new emphasis on use of benchmarks and estimates would make a difference. Right now, our sixth- and seventh-grade teachers are teaching the same concepts, and we hope they will not have to in the near future!

Additional References for Research and Teaching Implications

American Association for the Advancement of Science. (1993). *Benchmarks for science literacy.* New York: Oxford University Press. (pp. 288–291, 358–359).

National Council of Teachers of Mathematics. (2000). *Principles and standards for school mathematics.* Reston, VA: Author. (pp. 32–36, 214–221).

National Council of Teachers of Mathematics. (2003). *Research companion to principles and standards for school mathematics.* Reston, VA: Author. (pp. 95–110).

Van de Walle, J. A. (2007). *Elementary and middle school mathematics* (6th ed.). Boston: Pearson. (pp. 298–300, 303–308).

Curriculum Topic Study and Uncovering Student Thinking

What Is Your Estimate?

Keeley, P., & Rose, C. (2006). *Mathematics curriculum topic study: Bridging the gap between standards and practice.* Thousand Oaks, CA: Corwin. (Fractions, p. 121).

Related Elementary Probe:

Rose, C., Minton, L., & Arline, C. (2007). *Uncovering student thinking in mathematics: 25 formative assessment probes.* Thousand Oaks, CA: Corwin. (Comparing Fractions, p. 60).

Student Responses to What Is Your Estimate?

Sample Responses: A, d; B, a; and C, b

Student 1: "A: Since both are really close to one (see my picture), then the best answer would be 1 + 1 = 2. B: Since both are very far from one (see my picture) the best answer would be 0 + 0 = 0, but there is not a 0. $\frac{1}{4}$ was the smallest answer, so I picked it. C: Since $\frac{4}{5}$ is close to 1 and I know there is a $\frac{1}{2}$, I used $1 - \frac{1}{2} = \frac{1}{2}$ to choose the answer."

Sample Responses: Item A Distracters

Student 2: "F, 21. 13 and 8 is 21."

Student 3: "C,1. Both are less than 1, so they would add up close to 1."

Sample Responses: Item B Distracters

Student 4: "E, 2. 1 + 1 is 2."

Student 5: "G, 20. 1 and 1 and 8 and 10 is 20."

Sample Responses: Item C Distracters

Student 6: "E, 0. Both numbers are close to the same."

Student 7: "A, $\frac{1}{4}$. Fractions are small, so I picked the smallest answer."

6

Measurement, Geometry, and Data Probes

Quadrilaterals

Figure 6.1 Chapter 6 Probes

Grade-Span Bar Key		
	Target for Instruction Depending on Local Standards	
	Prerequisite Concept and Field Testing Indicate Students May Have Difficulty	

Question	Probe	Grade Span					
		K	1	2	3	4	5
Measurement and Geometry							
Do students understand the properties and characteristics of quadrilaterals?	Quadrilaterals (page 166)				3–4		5
Are students able to determine area without the typical length by width labelling?	What's the Area? (page 174)				3–5		
Do students pay attention to starting point when measuring with nonstandard units?	What's the Measure? (page 180)				3–5		
Data							
Are students able to choose the correct graphical representation when given a mathematical situation?	Graph Choices (page 187)				3–5		
Do students understand ways the median is affected by changes to a data set?	The Median (page 194)				3–5		

PROBE 21: QUADRILATERALS

Circle the name or names of the people you agree with.

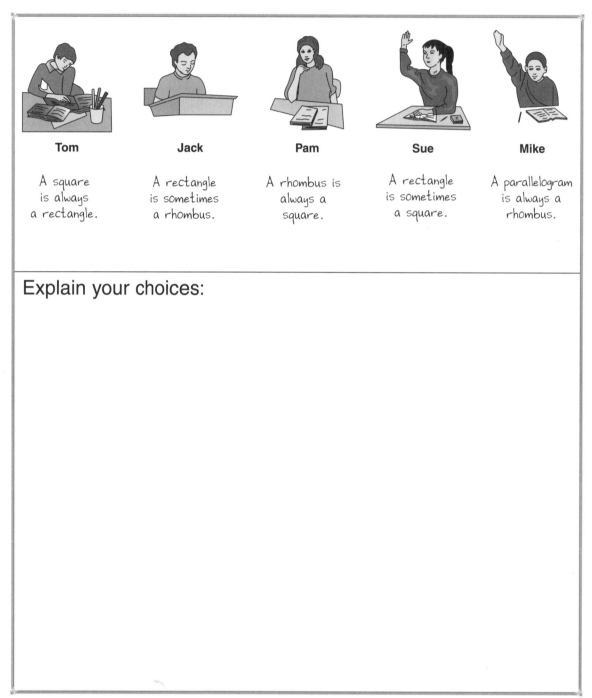

Tom	Jack	Pam	Sue	Mike
A square is always a rectangle.	A rectangle is sometimes a rhombus.	A rhombus is always a square.	A rectangle is sometimes a square.	A parallelogram is always a rhombus.

Explain your choices:

TEACHER NOTES: QUADRILATERALS

Questioning for Student Understanding

Do students understand the properties and characteristics of special types of quadrilaterals?

K–2	3–5

Uncovering Understanding

Quadrilaterals Content Standard: Geometry and Measurement

Variations/Adaptations:

- o Name That Shape
- o Variation: Is It a Polygon?
- o Variation: Is It a Circle?

Examining Student Work

The distracters may reveal *common misunderstandings* regarding geometric shapes, such as a lack of understanding of the various characteristics of classification of quadrilaterals.

- The correct statements are Tom, Jack, and Sue: Students who choose each of the correct statements understand the hierarchical nature of the classification of quadrilaterals, and their explanations will often include pictures to demonstrate examples and counterexamples as a means for justification. (See Student Responses 1 and 2.)
- Inclusion of Pam: Students who choose Pam typically do not understand that a rhombus does not have to include 90 degree angles and therefore a square is a special type of rhombus. (See Student Responses 3 and 4.)
- Inclusion of Mike: Students who choose Mike typically do not understand that a rhombus is a special type of parallelogram (4 equal-length sides). (See Student Response 5.)

Seeking Links to Cognitive Research

In grades 3–5: In the early grades, students will have classified and sorted geometric objects such as triangles or cylinders by noting general characteristics. Students in grades 3–5 should develop more-precise ways to describe shapes, focusing on identifying and describing the shape's properties and learning specialized vocabulary associated with these shapes and properties. To consolidate their ideas, students should draw and construct shapes, compare and discuss their attributes, classify them, and develop and consider definitions on the basis of a shape's properties, such as that a rectangle has four straight sides and four square corner. (National Council of Teachers of Mathematics [NCTM], 2000, p. 161)

Students advance through levels of thought in geometry. Van Hiele has characterized them as visual, descriptive, abstract/relational, and formal deduction (Van Hiele, 1986; Clements & Battista, 1992). At the first level, students identify shapes and figures according to their concrete examples. At the second level, students identify shapes according to their properties, and here a student might think of a rhombus as a figure with four equal sides. At the third level, students can identify relationships between classes of figures (e.g., a square is a rectangle) and can discover properties of classes of figures by simple logical deduction. At the fourth level, students can produce a short sequence of statements to logically justify a conclusion and can understand that deduction is the method of establishing geometric truth. (American Association for the Advancement of Science [AAAS], 1993, p. 352)

No single test exists to pigeonhole students at a certain level [see Van Hiele levels above]. At the upper elementary grades students should be pushed from level 1 to level 2. If students are not able to follow logical arguments or are not comfortable with conjectures and if-then reasoning, these students are likely still at level 1 or below. (Van de Walle, 2007, p. 414)

Students should carefully examine the feature of shapes in order to define and describe fundamental shapes, such as special types of quadrilaterals, and to identify relationships among the types of shapes. (NCTM, 2000, p. 233)

_T_eaching Implications

In order to support a deeper understanding for students in elementary school in regard to rectangles, the following are ideas and questions to consider in conjunction with the research.

Focus Through Instruction

- Students should be provided with materials and structured opportunities to explore shapes and their attributes.
- Students should analyze characteristics and properties of two- and three-dimensional shapes.
- Students should sort quadrilaterals by looking at examples and nonexamples of special types.
- Students should engage in mathematical conjectures about geometric relationships, such as why a square is a rectangle, but a rectangle is not always a square.
- Students need to develop more-precise ways to describe shapes using mathematics vocabulary associated with trapezoids, parallelograms, rectangles, rhombi, and squares.
- Use interactive technology to have students sort shapes by various attributes.

- View an example applet from NCTM's Illumination's (2000–2010) at http://illuminations.nctm.org/ActivityDetail.aspx?ID=34 (see information about using interactive applets in Chapter 1, page 22).

Questions to Consider
(when working with students as they grapple with the idea of special types of quadrilaterals)

- Do students begin to identify shapes by using the geometric attributes or properties of those shapes?
- Are students using mathematical language to describe parallelograms and address the similarities and differences of rectangles, rhombi, and squares?
- Are students able to describe the hierarchical nature of the classification of various quadrilaterals?
- Do students understand that one counterexample proves a statement is false?

Teacher Sound Bite

I was pleased with my students understanding of the relationship between squares and rectangles but clearly saw in the student responses that parallelograms and rhombi are still a bit elusive. I think they don't have as much experience with considering all types of quadrilaterals and how each relates to the others. To help with this, I created a hands-on activity where students had two sets of 20 different quadrilaterals (one set yellow and the other set blue). First, I gave students the names of two shapes and asked students to find examples of the first shape using the yellow examples and the second shape using the blue examples. I then asked students to compare the examples, noticing those common to both shapes. I continued the activity several times while writing various statements the board. The activity seemed to have made a difference with students understanding of the relationship among quadrilaterals.

Additional References for Research and Teaching Implications

National Council of Teachers of Mathematics. (2000). *Principles and standards for school mathematics.* Reston, VA: Author. (p. 165).

National Council of Teachers of Mathematics. (1993). *Research ideas for the classroom: Early childhood mathematics.* New York: Macmillan. (pp. 204–220).

National Research Council. (2001). *Adding it up: Helping children learn mathematics.* Washington, DC: National Academy Press. (pp. 284–285).

Curriculum Topic Study and Uncovering Student Thinking

Quadrilaterals

Keeley, P., & Rose, C. (2006). *Mathematics curriculum topic study: Bridging the gap between standards and practice.* Thousand Oaks, CA: Corwin. (Quadrilaterals, p. 161).

Related Probes:

Rose, C., Minton, L., & Arline, C. (2007). *Uncovering student thinking in mathematics: 25 formative assessment probes.* Thousand Oaks, CA: Corwin. (What Does a Rectangle Look Like? p. 141; Parallelograms, p. 163).

Student Responses to Quadrilaterals

Sample Responses: Tom, Jack, and Sue

Student 1: "Tom and Sue were easy because they are related. A square has to be a rectangle—the only difference is that for a square the sides are all the same. So, if the sides aren't the same, it is a rectangle with four 90-degree corners and four sides that never would cross if you make them longer. Since a rhombus is sometimes a square that means if a rectangle is a square then it is a rhombus too. This only happens for some rectangles though, so Jack saying *sometimes* would be correct."

Sample Responses: Tom and Sue, Jack Excluded

Student 2: "We talk about squares and rectangles all the time, so I know Tom and Sue are right in what they are saying."

Sample Responses: Inclusion of Pam

Student 3: "I also picked Pam because a rhombus is just a sidewise square."

Student 4: "I do agree with Pam because I learned that there are lots of names for four-sided figures, and they connect to each other."

Sample Responses: Inclusion of Mike

Student 5: "For why I circled Mike—a rhombus and parallelogram are the same thing because they both have four sides, and opposite sides parallel."

PROBE 21A: VARIATION: NAME THAT SHAPE

Circle all correct names for each shape.	Explain your thinking:
1. quadrilateral square rectangle rhombus parallelogram trapezoid	
2. quadrilateral square rectangle rhombus parallelogram trapezoid	
3. quadrilateral square rectangle rhombus parallelogram trapezoid	

21a

PROBE 21B: VARIATION: IS IT A POLYGON?

Circle only the figures that are polygons.

A)

B)

C)

D)

E)

F)

Explain your choices:

PROBE 21C: VARIATION: IS IT A CIRCLE?

Circle only the figures that are circles.

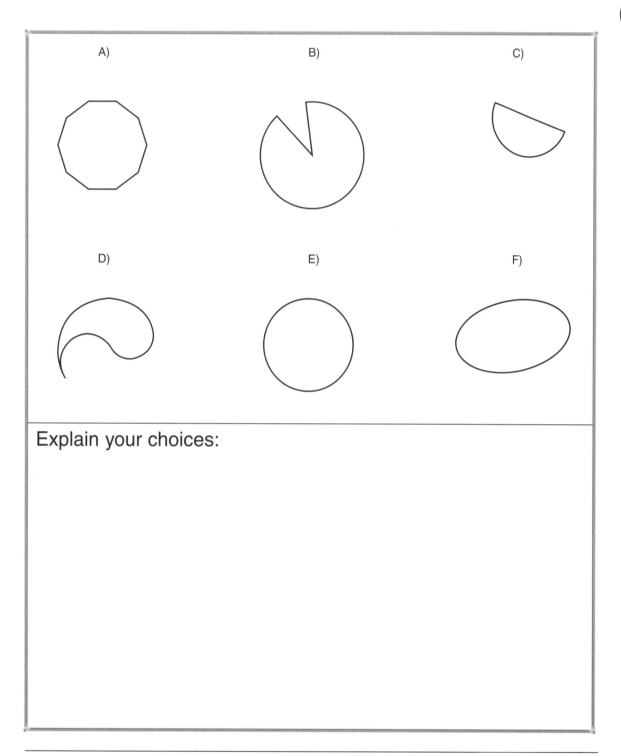

A)

B)

C)

D)

E)

F)

Explain your choices:

PROBE 22: WHAT'S THE AREA?

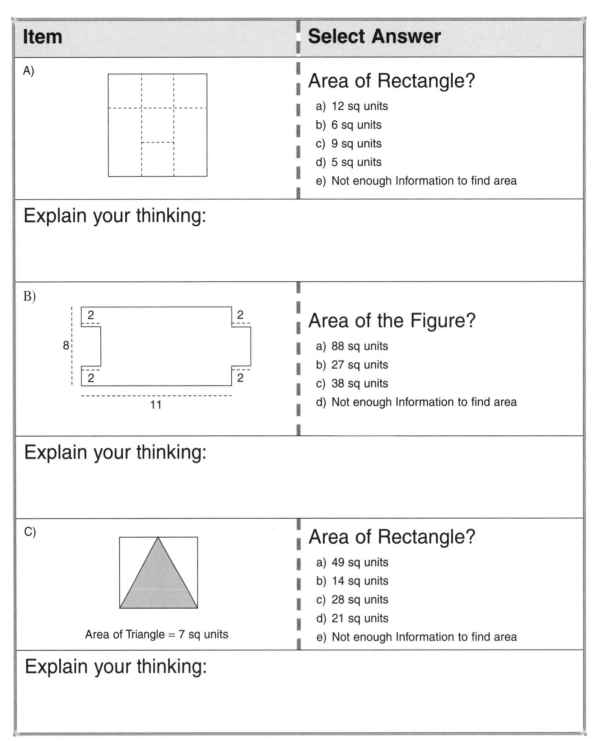

Item	Select Answer
A)	**Area of Rectangle?**
	a) 12 sq units
	b) 6 sq units
	c) 9 sq units
	d) 5 sq units
	e) Not enough Information to find area
Explain your thinking:	
B)	**Area of the Figure?**
	a) 88 sq units
	b) 27 sq units
	c) 38 sq units
	d) Not enough Information to find area
Explain your thinking:	
C)	**Area of Rectangle?**
Area of Triangle = 7 sq units	a) 49 sq units
	b) 14 sq units
	c) 28 sq units
	d) 21 sq units
	e) Not enough Information to find area
Explain your thinking:	

TEACHER NOTES: WHAT'S THE AREA?

Questioning for Student Understanding

Are students able to determine area without the typical length by width labeling?

K–2	3–5

Uncovering Understanding

What is the Area? Content Standard: Geometry and Measurement

Variation/Adaptation:

o Variation: What's the Area?

Examining Student Work

The distracters may reveal *common misunderstandings* regarding geometric measurement such as a lack of *conceptual understanding* of the similarities and differences between area and perimeter and not being able to decipher the length and width when not labeled directly.

- The correct answers are A, c; B, a; and C, b: Students who answer each of these not only correctly understand area as length times width but are also able to calculate the area when a portion of the rectangle has been translated, and they understand the relationship between the area of a triangle and rectangle. (See Student Response 1.)
- Distracters Item A: Students who answer *a, 12 sq units* are typically confusing area with perimeter by determining and counting the number of unit lengths on each side of the rectangle. Students who answer *b, 7 sq units* typically are counting both the unit squares and rectangles (1 each). Students who answer *d, 5 sq units* typically are counting the units squares shown within the rectangle. Students who answer *e, not enough information* typically believe the length and width cannot be determined from the information provided (relying on the *1* and *w* being labeled or all unit squares showing). (See Student Responses 2 and 3.)
- Distracters Item B: Students who answer *b, 27 sq units* are typically confusing adding all of the side lengths provided with finding the area. Students who answer *c, 38 sq units* are typically confusing area with perimeter. Students who answer *d, not enough information* typically believe the area cannot be determined from the information provided because a portion of the original rectangle as been translated (they may be having difficulty with conservation of area). (See Student Responses 4 and 5.)
- Distracters Item C: Students who answer *a, 49 sq units* are typically confusing the area of the triangle with the length of a side of the rectangle and

assume the rectangle shown is a square. Students who answer *c, 28 sq units* are typically confusing the area of the triangle with the length of a side of the rectangle and assume the rectangle shown is a square, but then they also confuse finding the area of the square with finding the perimeter. Students who answer *d, 21 sq units* typically confuse the area of the triangle with the length of the triangle and use this information to calculate the perimeter of the triangle. Students who answer *e, not enough information* typically believe the length and width cannot be determined from the information provided (they rely on the *1* and *w* being labeled and don't consider the relationship between the area of the triangle and rectangle provided). (See Student Responses 6 and 7.)

<u>S</u>eeking Links to Cognitive Research

Students have performed poorly on NAEP items targeting the following measurement topics: choosing the correct numerical expression for the area of a given geometric figure, determining the number of square tiles needed to cover a region of given dimensions, determining the number of boxes of square tiles needed to cover a region of given dimensions, showing a number of different ways a region can be divided to find the area, and determining the surface area of a rectangular solid. (NCTM, 2006, p. 126–127)

Although students can often recall standard formulas for finding areas of squares and rectangles, other aspects of area measure remain problematic. . . . Research suggests that many students in elementary school do not "see" the product as a measurement. Even students with experience using square tiles can view the tiles as things to be counted rather than the subdivision of a plane. (NCTM, 2003, p. 185)

Area and perimeter are continually a source of confusion for students. Perhaps it is because both involve regions to be measured or because students are taught formulas for both concepts and tend to get formulas confused. (Van de Walle, 2007, p. 386)

<u>T</u>eaching Implications

In order to support a deeper understanding for elementary students in regard to geometric measurement, the following are ideas and questions to consider in conjunction with the research.

Focus Through Instruction

- Students should be provided opportunities to experiment with area using concrete objects and various shapes.
- Composing and decomposing shapes should be used as a method of finding areas for various two-dimensional objects, including squares and rectangles.
- Students should develop formulas through inquiry and investigation.

- Students should understand how the length and/or the width of a rectangle correspond with its base and height.
- Students should be able to apply area concepts beyond simply calculating an answer when given a base and height.

Questions to Consider (when working with students as they grapple with the idea of area and perimeter)

- Are students able to distinguish correctly between perimeter and area?
- Do the students understand that area is conserved when decomposing and recomposing a shape?
- Do students interpret the results of area calculations as a measure, understanding and applying appropriate units?
- Do students understand the relationship of the area of a triangle to a corresponding rectangle (with the same base and height)?
- Are students able to determine the appropriate measures needed in order to calculate the area of the shape?

Teacher Sound Bite

My students tend to *choose not enough info* whenever they come across a problem that looks different than one they have seen in the past. Because I don't want them to think the answer never is a correct choice, I created two items to use as the problem of the day where sometimes there really isn't enough information given. As for the application of area concepts, about half of the students who did not choose *not enough info* chose at least one wrong answer because of faulty reasoning. The most common mistake was finding the perimeter for B. I think, once students see all of those numbers, they automatically decide to add them all together.

Additional References for Research and Teaching Implications

National Council of Teachers of Mathematics. (2000). *Principles and standards for school mathematics.* Reston, VA: Author. (pp. 173, 244).

National Council of Teachers of Mathematics. (2003). *Research companion to principles and standards for school mathematics.* Reston, VA: Author. (pp. 180–186).

Stepans, J. I., Schmidt, D. L., Welsh, K. M., Reins, K. J., Saigo, B. W. (2005). *Teaching for K–12 mathematical understanding using the conceptual change model.* St. Cloud, MN: Saiwood. (pp. 233–235).

Van de Walle, J. A. (2007). *Elementary and middle school mathematics* (6th ed.). Boston: Pearson. (pp. 382–386).

Curriculum Topic Study and Uncovering Student Thinking

What's the Area?

Keeley, P., & Rose, C. (2006). *Mathematics curriculum topic study: Bridging the gap between standards and practice.* Thousand Oaks, CA: Corwin. (Perimeter, Area, and Volume, p. 175).

Related Elementary Probes:

Rose, C., Minton, L., & Arline, C. (2007). *Uncovering student thinking in mathematics: 25 formative assessment probes.* Thousand Oaks, CA: Corwin. (Fractional Parts, p. 49; Volume of a Box, p. 168; Area of a Figure, p. 172).

Student Responses to What's the Area?

Sample Responses: A, c; B, a; and C, b

Student 1: "A: Just draw in the missing lines, and it's a 3 = 3 box. To find area, multiply the two numbers, so the answer is 9. B: Cut of one the extra flaps and glue it into the empty space at the beginning. You just need two numbers to multiply for the answer, so use 8 times 11, and the answer is 88. C: If you cut up the two white triangles and flip them, they would fit into the grey one, so it is like there are two of those size triangles. If one is 7, then 7 and 7 would be an answer of 14."

Sample Responses: Incorrect Responses for Item A

Student 2: "A: I drew in the rest of the dotted lines. I counted the number of divisions on each side then I add them all together."

Student 3: "B: I counted all the pieces inside."

Sample Responses: Incorrect Responses for Item B

Student 4: "C: I added up all the numbers plus 11 more for the top, which is the same length as the bottom."

Student 5: "D: Some of the lines don't have numbers, so you can't do anything."

Sample Responses: Incorrect Responses for Item C

Student 6: "A: multiply l by w, so 7×7 is 49."

Student 7: "E: There is no way to figure out the measures if all you know is the area of triangle. The bottom and height could be anything that makes 7."

PROBE 22A: VARIATION: WHAT'S THE AREA?

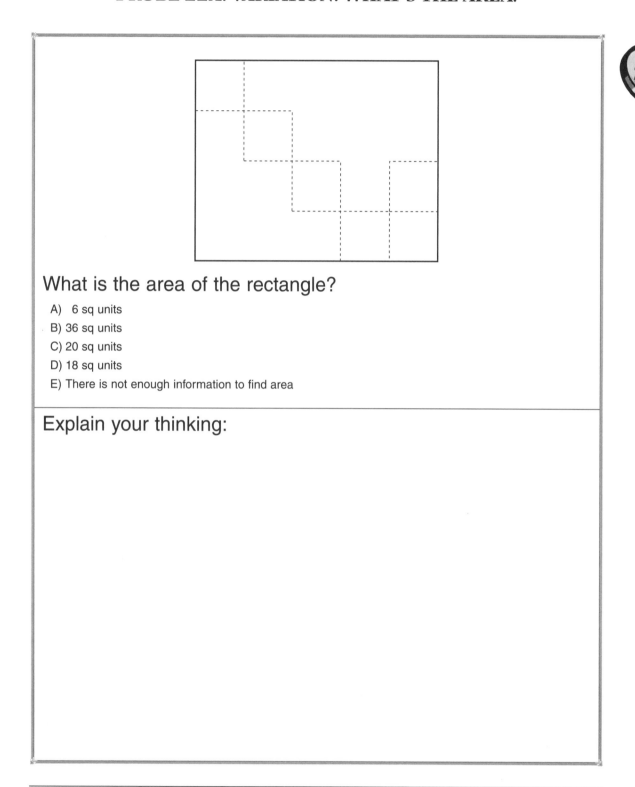

What is the area of the rectangle?

A) 6 sq units

B) 36 sq units

C) 20 sq units

D) 18 sq units

E) There is not enough information to find area

Explain your thinking:

PROBE 23: WHAT'S THE MEASURE?

A)

Is this line approximately $2\frac{1}{4}$ units long?_____

Explain why or why not:

B)

Is this line approximately $2\frac{1}{4}$ units long?_____

Explain why or why not:

C)

Is this line approximately $2\frac{1}{4}$ units long?_____

Explain why or why not:

D)

Is this line approximately $2\frac{1}{4}$ units long?_____

Explain why or why not:

TEACHER NOTES: WHAT'S THE MEASURE?

Questioning for Student Understanding

Are students able to choose the correct measure of a line given a change in the interval or given a nonzero starting point?

K–2	3–5
	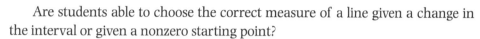

Uncovering Understanding

What's the Measure? Content Standard: Measurement

Examining Student Work

The distracters may reveal *common errors* made when measuring or a lack of understanding when measuring parts of a unit.

- The correct answers are A, no; B, yes; C, yes; and D, yes: Students who correctly choose these answers are taking into consideration the starting point and are able to understand one-fourth given various interval choices. (See Student Responses 1 and 3.)
- Distracter A: Students who *include* A typically do not consider the nonzero starting point and give the length as the number on the ruler aligned to the endpoint of the segment. (See Student Response 2.)
- Distracter C: Students who *exclude* C are not considering that each interval represents $\frac{1}{10}$, with the length of the line approximately $2\frac{1}{2}$ units long. (See Student Response 4.)
- Distracter D: Students who *exclude* D are not considering the nonzero starting point. (See Student Response 5.)

Seeking Links to Cognitive Research

Length is usually the first attribute students learn to measure. Be aware, however, that length measurement is not immediately understood by young children. . . . The temptation is to explain to students how to use units to measure and then send them off to practice measuring. This approach will shift attention to the procedure and away from developing an understanding of measurement using units. (Van de Walle, 2007, p. 379)

Children's understanding of zero-point is particularly tenuous. Only a minority of young children understand that any point on a scale can serve as the starting point, and even a significant minority of older children (e.g., fifth grade) respond to non-zero origins by simply reading off whatever number on a ruler aligns with the end of the object (Lehrer et al., 1998a). Many children throughout schooling begin measuring with one rather than zero (Ellis, Siegler, & Van Voorhis, 2001). (NCTM, 2003, p. 183)

Conceptual understanding of measurement includes, among others, understanding of iteration and understanding of the origin. Specific ideas include 1) the need for identical units; 2) partitioning of a unit; 3) conformity on the scale [meaning] that any location on the scale can serve as the origin; 4) mental coordination of the origin and the endpoint of the scale and the resulting measure. Most studies suggest that these understandings of units of length are acquired over the course of the elementary grades. (NCTM, 2003, p. 182)

When using conventional tools such as rulers and tape measures for measuring length, students will need instruction to learn to use these tools properly. For example, they will need to recognize and understand the markings on a ruler, including where the "0," or beginning point, is located. (NCTM, 2000, p. 173)

Teaching Implications

In order to support a deeper understanding for students in regard to measurement, the following are ideas and questions to consider in conjunction with the research.

Focus Through Instruction

- By constructing their own measurement tools, students can discover important measurement ideas, including the partitioning of a unit into smaller units to accommodate more-precise length measures.
- An understanding of measurement can serve as a basis for work with fraction and decimal numbers and the idea of scales on a coordinate graph.
- Students need experience with a variety of standard and nonstandard measuring tools with varying intervals.
- Opportunities to measure length with "broken" rulers can help students learn to compensate for a nonzero starting point.
- Discussion around precision versus accuracy can help students approximate measures of objects that do not match up exactly with marks on a ruler.
- Use interactive technology to have students practice measuring lengths.
- View an example measurement-related applets at http://maine.edc .org/file.php/1/K6.html (see information about using interactive applets in Chapter 1, page 22).

Questions to Consider (when working with students as they grapple with the idea of measuring length)

- Do students understand length measure as the result of matching a length with a number of units rather than as a number on the ruler?
- Do students realize the size of the unit determines the number of units that make up the length of an object?
- When measuring with a "broken" ruler, do students compensate by translating the object and considering the difference from a zero starting point?

Teacher Sound Bite

Year after year, students have difficulty with correctly using a ruler to measure. I know considering the starting point is a main issue, and I've have always instructed students to pay attention and use various *broken* rulers in an activity to force students into thinking about what to use as a starting point and how to adjust the final measure accordingly. This year, I used the probe prior to these activities and posted a double-line graph of responses (the number of yes or no answers for each item). As we moved through the activities, I periodically referred back to items and polled the students for answers. Many students, recognizing they had incorrect answers, were actually anxious to get their original responses back, so they could change their answers.

Additional References for Research and Teaching Implications

Bay Area Mathematics Task Force. (1999). *A mathematics sourcebook for elementary and middle school teachers.* Novato, CA: Arena Press. (pp. 27–35).

National Council of Teachers of Mathematics. (2000). *Principles and standards for school mathematics.* Reston, VA: Author. (pp. 171–174, 243).

National Council of Teachers of Mathematics. (2003). *Research companion to principles and standards for school mathematics.* Reston, VA: Author. (pp. 180–184).

National Research Council. (2001). *Adding it up: Helping children learn mathematics.* Washington, DC: National Academy Press. (pp. 281–282).

Van de Walle, J. A. (2007). *Elementary and middle school mathematics* (6th ed.). Boston: Pearson. (pp. 374–382).

Curriculum Topic Study and Uncovering Student Thinking

What's the Measure?

Keeley, P., & Rose, C. (2006). *Mathematics curriculum topic study: Bridging the gap between standards and practice.* Thousand Oaks, CA: Corwin. (Length, p. 171).

Related Probes:

Rose, C., Minton, L., & Arline, C. (2007). *Uncovering student thinking in mathematics: 25 formative assessment probes.* Thousand Oaks, CA: Corwin. (How Long is the Pencil, p. 146).

Rose, C., & Arline, C. (2009). *Uncovering student thinking in mathematics, grades 6–12: 30 formative assessment probes for the secondary classroom.* Thousand Oaks, CA: Corwin. (What's the Measure, p. 107; Variation: What's the Measure? p. 111).

Student Responses to What's the Measure?

> ### Sample Responses: A, no; B, yes; C, yes; and D, yes
>
> Student 1: "I think that you count two full units then count the little lines, or make your own, to check if there is $\frac{1}{4}$ of another unit."

> ### Sample Responses: Inclusion of A
>
> Student 2: "This is the only ruler that looks right. The line goes past the two to the first longer line."

> ### Sample Responses: Inclusion of B
>
> Student 3: "Yes, this ruler has unit marks on it, so it is easy to read it. You just have to add some more parts."

> ### Sample Responses: Exclusion of C
>
> Student 4: "This is almost 2 and $\frac{1}{4}$, but I said no because the line should really go to the fourth mark after the 2 mark."

> ### Sample Responses: Exclusion of D
>
> Student 5: "This is $\frac{1}{4}$th, but it's $3\frac{1}{4}$ not $2\frac{1}{4}$."

PROBE 23A: VARIATION: LENGTH OF LINE

About how long is the line?

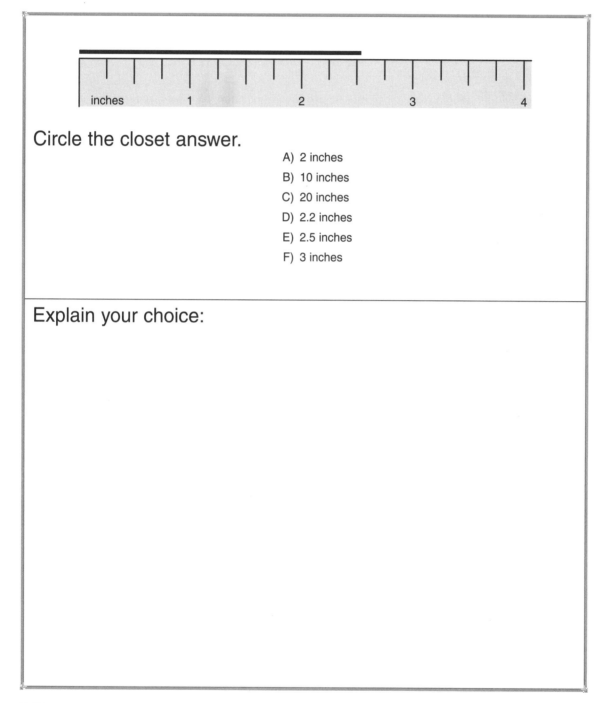

Circle the closet answer.

- A) 2 inches
- B) 10 inches
- C) 20 inches
- D) 2.2 inches
- E) 2.5 inches
- F) 3 inches

Explain your choice:

PROBE 23B: VARIATION: WHAT'S THE MEASURE?

Circle only the lines that are about 2 units long.

A)

B)

C)

D)

E)

Explain your choices:

PROBE 24: GRAPH CHOICES

The Number of Students Who Walked to School in the Last Five Days

Given only the title and the portions of the graphs shown below, which of the following could *possibly* be a *correct graph* of the situation?

24

Graph A)

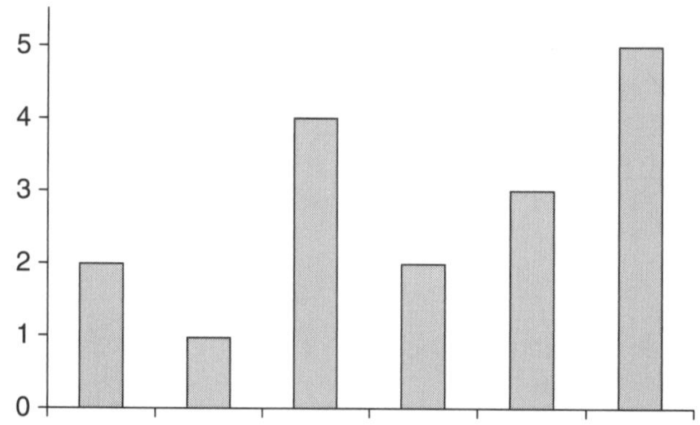

Possible graph? Yes   No

Explain your choice:

Graph B)

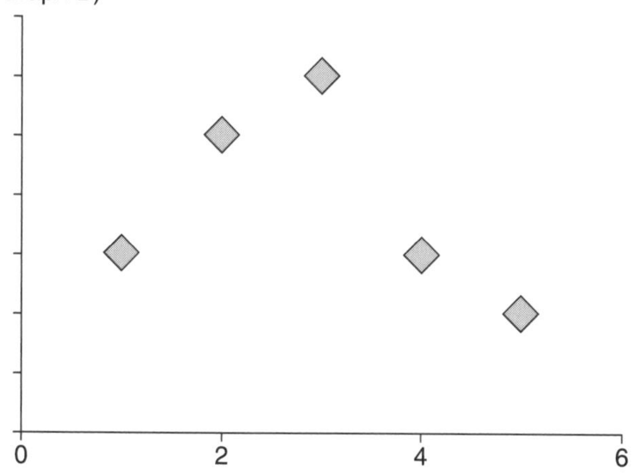

Possible graph? Yes   No

Explain your choice:

Graph C)

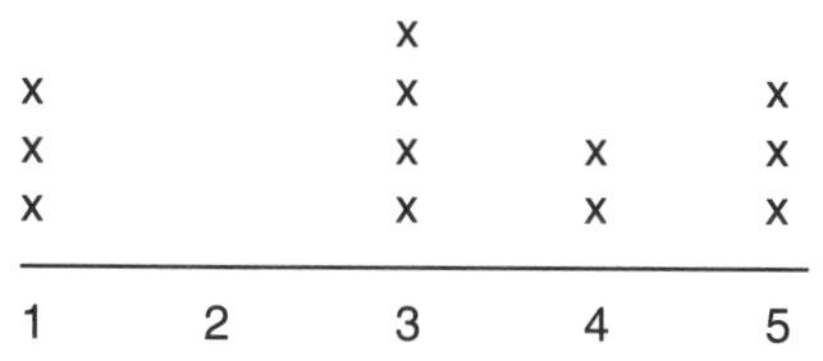

		x		
x		x		x
x		x	x	x
x		x	x	x
1	2	3	4	5

Possible graph? Yes   No

Explain your choice:

TEACHER NOTES: GRAPH CHOICES

Questioning for Student Understanding

Are students able to choose an appropriate graphical representation when given a numerical situation?

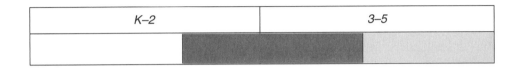

Uncovering Understanding

Graph Choices Content Standard: Data Analysis

Examining Student Work

The distracter in this probe may reveal *common errors* in choosing appropriate representations when given a categorical situation.

- The correct answers are B and C: Given the "last five days," the graph should not include more than five days worth of data (it can be less than five though if zero students walked to school on any given day). There are no restrictions to the maximum or minimum number of students (the number of students per day could be zero or greater). Although the dot plot in Graph C does not include data for Day 2, this representation is still a correct choice for a situation where zero students walked to school on Day 2, and if each X represents a certain number of students. Watch for how students justify their responses, as a correct response may still indicate an overgeneralization such as the X represents one student. (See Student Responses 2, 3, and 5.)
- Distracter A: Some students will choose this example because they believe the 5 on the vertical axis represents the five days in the scenario. (See Student Response 1.)
- Distracter, Answering No to B or C: Students who do not choose B and/or C cannot make inferences about a mathematical situation from the graphic representation provided. These students typically believe that the numbers 0–5 or 1–5 need to be labeled or that each day needs a nonzero value. (See Student Responses 4.)

Seeking Links to Cognitive Research

Children with little experience with the various methods of picturing data will not be aware of the many options that are available. (Van de Walle, 2007, p. 459)

Students read graphs point-by-point and ignore their global features. This has been attributed to algebra lessons where students are given questions that they could easily answer from a table of ordered pairs. (AAAS, 1993, p. 351)

A new view of mathematical representations in general and graphing in particular has slowly emerged in the past decade. Instead of being isolated curricular items to be taught and tested as ends in themselves, graphs along with diagrams, charts, number sentences, and formulas, are increasingly seen as "useful tools for building understanding and for communicating both information and understanding" (NCTM, 2000). As such, graphs and other representations have come to play an increasingly important role in mathematical activities in school. (NCTM, 2003, p. 250)

Students need multiple experience viewing data from different perspectives. . . . Dot plots are hard for novices (Cobb, 1999), because they associate the height of a stack with the magnitude of attributes rather than with frequencies. (Stepans et al., 2007, p. 193)

Little is known about how graphic skills are learned and how graph production is related to graph interpretation. Microcomputer-Based Laboratories (MBLs) are known to improve the development of students' abilities to interpret graphs. (AAAS, 1993, p. 351)

Teaching Implications

In order to support a deeper understanding for students in regard to graphical interpretation, the following are ideas and questions to consider in conjunction with the research.

Focus Through Instruction

- Have students analyze a variety of graphs that were created by other sources: students, newspapers, the Internet, and so on.
- Have students work with several graphs simultaneously. Have them compare and contrast them. Have them write a heading (or description) for each graph.
- Have students verbalize what the horizontal and vertical axes are representing.
- Provide multiple examples and nonexamples of graphic representations for situations where students do not have access to the actual data.
- Involve students in making decisions about how to represent data.

Questions to Consider (when working with students as they grapple with the idea of graphical representation)

- Do students have an understanding of the categories involved in a given situation in order to make decisions about horizontal and vertical intervals and categories?

- Do students understand global features of a graph when the graph is missing labels and/or is presented without actual data points?
- Have the students been given multiple opportunities to collect their own data and graph the data using a variety of appropriate graph types?
- Can students use a variety of graphic representations when given a data set?
- Are students only creating graphs, or are they also analyzing graphs from other sources?

Teacher Sound Bite

The first time I gave this probe, my students were very confused as they had never been asked to look at multiple forms of graphs without actually having the data to create the graphs themselves. I had to seriously consider whether this type of graphing had a place in my fifth-grade classroom, especially as I would have to supplement the math program in order to do so. I finally decided that I would address it by discussing the probe in math workshop and then by continuing to create additional scenarios and graphs to use in workshop times during a 1-month period. Next year, I plan to incorporate this during the graph unit, which this year's students had already completed before I used the probe.

Curriculum Topic Study and Uncovering Student Thinking

Graph Choices

Keeley, P., & Rose, C. (2006). *Mathematics curriculum topic study: Bridging the gap between standards and practice.* Thousand Oaks, CA: Corwin. (Line Graphs, Bar Graphs, and Histograms, p. 179; Graphic Representation, p. 196).

Related Probes:

Rose, C., Minton, L., & Arline, C. (2007). *Uncovering student thinking in mathematics: 25 formative assessment probes.* Thousand Oaks, CA: Corwin. (Name of the Graph, p. 145; Graph Construction, p. 151).

Additional References for Research and Teaching Implications

American Association for the Advancement of Science. (1993). *Benchmarks for science literacy.* New York: Oxford University Press. (pp. 297, 351).

National Council of Teachers of Mathematics. (2000). *Principles and standards for school mathematics.* Reston, VA: Author. (pp. 49–50, 176–180, 248–253).

National Council of Teachers of Mathematics. (2003). *Research companion to principles and standards for school mathematics.* Reston, VA: Author. (pp. 202, 250–260).

Stepans, J. I., Schmidt, D. L., Welsh, K. M., Reins, K. J., Saigo, B. W. (2005). *Teaching for K–12 mathematical understanding using the conceptual change model.* St. Cloud, MN: Saiwood. (pp. 189–230).

Van de Walle, J. A. (2007). *Elementary and middle school mathematics* (6th ed.). Boston: Pearson. (pp. 452–464).

Student Responses to Graph Choices

Sample Responses: Graph A

Student 1: "Yes. There are five days and the graph shows from 0 to 5."

Student 2: "No. This would show six days because of the 6 bars. There are too many days on the graph."

Sample Responses: Graph B

Student 3: "Yes. There are five dots which show the number students for each day. Monday is diamond one because it is in line with 1; Tuesday is diamond two because it is in line with 2; Wednesday is diamond three because it is in line with 3; Thursday is diamond four because it is in line with 4; and, Friday is diamond five because it is in line with 5."

Student 4: "No. There are no numbers on the vertical part, and I wouldn't use this kind of chart anyway."

Sample Responses: Graph C

Student 5: "Yes. This has right numbers if each X is counting one student, and the numbers listed are for each day. It is made correctly because the teacher said that you don't have to include things when there isn't a number for the spot."

Student 6: "No. I didn't choose this because it skips right over the number two. That is the wrong information."

PROBE 24A: VARIATION: NAME THE GRAPH

Bob made the graph below.

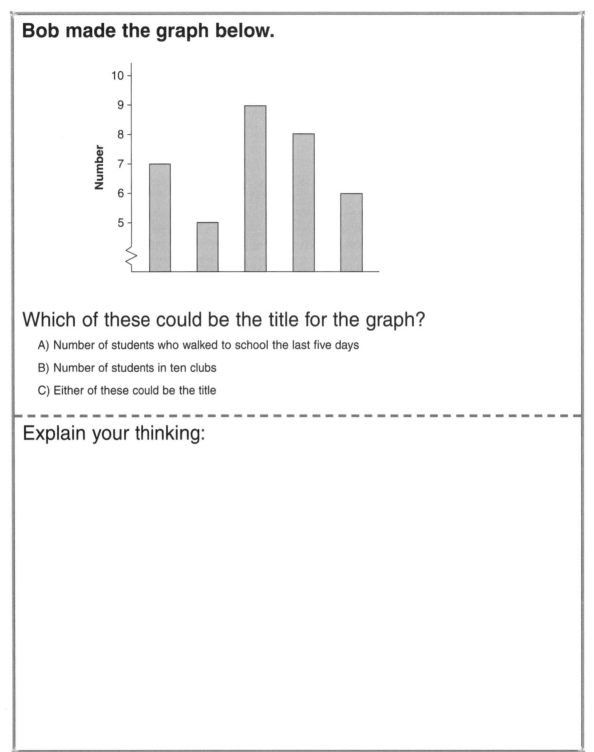

Which of these could be the title for the graph?

A) Number of students who walked to school the last five days

B) Number of students in ten clubs

C) Either of these could be the title

Explain your thinking:

PROBE 24B: VARIATION: WHAT DOES THE GRAPH SAY?

Each student in the class chose one favorite color, and a bar chart was created using the data.

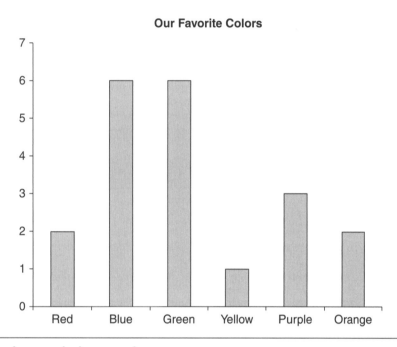

Circle the true statements.

A) Nobody chose *yellow*.

B) *Red* and *orange* were chosen the same number of times.

C) The numbers of students who chose *blue* is more than the number who chose *red* and *yellow* combined.

D) There are 18 students in the class.

E) More than half the class chose *blue* or *green*.

Explain your choices:

PROBE 25: THE MEDIAN

Find the median of each data set provided.

A) Data set: The number of students who walked to school each day this week.

Mon 2   Tues 6   Wed 7   Thus 3   Fri 2

Circle the median of the data set. | Explain your answer:

A) 2   D) 5
B) 3   E) 6
C) 4   F) 7

B) Data set: Thirteen students' number of pockets.

```
                                    X
                 X            X     X
          X      X            X     X
          X      X      X     X     X
          _____
          1      2      3     4     5
```

Circle the median of the data set. | Explain your answer:

A) 1   D) 4
B) 2   E) 5
C) 3   F) 13

TEACHER NOTES: THE MEDIAN

Questioning for Student Understanding

Do students understand the median and how it is affected by changes to a data set?

K–2	3–5

Uncovering Understanding

Finding the Median Content Standard: Data Analysis

Variation/Adaptation:

- Effect on the Median Card Sort

Examining Student Work

This probe may uncover a lack of *conceptual understanding* of median as a measure of center and *common errors* associated with not ordering the data set.

- The correct responses are A, 3 and B, 4. (See Student Responses 1 and 4.)
- Distracters for Data Set A: Students who answer *a, 2* are typically giving the mode. Students who answer *c, 4* are typically giving the mean. Students who answer *d, 5* are counting the number of data in the data set. Students who answer *e, 6* typically respond, "I don't know," or, "I'm not sure." Students who choose *f, 7* typically are viewing the seven as the middle of the data set without having ordered the numbers listed. (See Student Responses 2 and 3.)
- Distracters for Data Set B: Students who answer *a, 1* are typically giving the minimum value. Students who answer *b, 2* are typically guessing. Students who answer *c, 3* are viewing the three as the middle of the list of numbers without having considered the 13 data points. Students who answer *e, 5* are typically giving the mode. Students who answered *f, 13* are typically counting the number of data in the data set. (See Student Responses 5 and 6.)

Seeking Links to Cognitive Research

A study by Russell et al. (2002a) shows that young students ideas about averages are based on everyday meanings that draw on qualitative rather then quantitative notions of what is "typical." Teachers in the study describe their students as "heading straight for the mode" [when asked to describe the typical value of a data set]. (NCTM, 2003, p. 203)

Students should make distributions for many data sets, their own and published sets, which have already inspired some meaningful questions. The idea of a middle to a data set should be well motivated—say by asking for a simple way to compare two groups—and various kinds of middle should be considered. (AAAS, 1993, p. 228)

Much of students' work with data in grades 3–5 should involve comparing related data sets. Noting the similarities and differences between two data sets requires students to become more precise in their descriptions of the data. In this context, students gradually develop the idea of a "typical," or average, value. Building on their informal understanding of "the most" and "the middle," students can learn about three measures of center—mode, median, and, informally, the mean. Students need to learn more than simply how to identify the mode or median in a data set. They need to build an understanding of what, for example, the median tells them about the data, and they need to see this value in the context of other characteristics of the data. (NCTM, 2000, p. 179)

*T*eaching Implications

In order to support a deeper understanding for students in regard to the median, the following are ideas and questions to consider in conjunction with the research.

Focus Through Instruction

- Experiences with collecting, creating, and comparing data sets provide a powerful motivator to learning about how to find descriptive measures.
- Intuitive sense of middle from real-life experiences help students understand the process of finding the median but can also distract from understanding the necessity of ordering the numbers.

Questions to Consider (when working with students as they grapple with finding the median)

- Do students understand the difference between the mode, median, and mean?
- Are students flexible in their use of various measures of center and spread when describing and comparing data sets?
- Do students pay attention to the order of the numbers prior to finding the middle value?

Teacher Sound Bite

In the past, I haven't spent a lot of instructional time on finding the median because my students seem to pick it up rather quickly and always seem to have a much more difficult time with finding the mean. I was anticipating that students wouldn't have any trouble on the probe but found that close to half of my student intuitively provided the middle number as listed in the problem without even considering first ordering them from smallest to largest. These students were quick to realize the mistake on the first problem, but we spent a good deal more time on the second because of the extra step of creating the data list from the graph provided.

Additional References for Research and Teaching Implications

American Association for the Advancement of Science. (1993). *Benchmarks for science literacy.* New York: Oxford University Press. (pp. 228–229, 353–354).

National Council of Teachers of Mathematics. (2000). *Principles and standards for school mathematics.* Reston, VA: Author. (pp. 50, 176, 179–180, 248, 250–251, 327–328, 342–344).

National Council of Teachers of Mathematics. (2003). *Research companion to principles and standards for school mathematics.* Reston, VA: Author. (pp. 202–209).

Stepans, J. I., Schmidt, D. L., Welsh, K. M., Reins, K. J., & Saigo, B. W. (2005). *Teaching for K–12 mathematical understanding using the conceptual change model.* St. Cloud, MN: Saiwood. (pp. 189–230).

Van de Walle, J. A. (2007). *Elementary and middle school mathematics* (6th ed.). Boston: Pearson. (p. 464).

Curriculum Topic Study and Uncovering Student Thinking

The Median

Keeley, P., & Rose, C. (2006). *Mathematics curriculum topic study: Bridging the gap between standards and practice.* Thousand Oaks, CA: Corwin. (Measures of Center and Spread, p. 181).

Related Elementary Probe:

Rose, C., & Arline, C. (2009). *Uncovering student thinking in mathematics, grades 6–12: 30 formative assessment probes for the secondary classroom.* Thousand Oaks, CA: Corwin. (What Do You Mean? p. 137).

Student Responses to The Median

Sample Responses: Item A

Student 1: "B, 3. First, list the numbers form small to big. Next, cross out one number from each end until you reach one number left. This was a short list, so it was easy to see the middle one."

Student 2: "F, 7. Median means middle, so the middle of the list is 7."

Student 3: "A, 2. It is the only number listed more than once."

Sample Responses: Item B

Student 4: "D, 4. Each X means the number below the line, so the list of numbers looks like this—1, 1, 2, 2, 2, 3, 4, 4, 4, 5, 5, 5, 5. There are 13 numbers, so count 7 numbers in to find the middle."

Student 5: "C, 3. The numbers used in the chart are 1, 2, 3, 4, and 5, so the middle number is 3."

Student 6: "C, 3. I added the numbers 1 + 2 + 3 + 4 + 5 and got 15, then I divided by 5 to get 3."

PROBE 25A: VARIATION: EFFECT ON THE MEDIAN CARD SORT

Directions: Discuss the following data collection example with your students. Together, find the median.

For each of the following, ask the students to determine (without actually calculating) whether adding a new data point for Friday would impact the median of the set.

Data

Each day of the week, the class collects data on the number of students who walked to school. The following information has been collected so far:

Monday: 3 students
Tuesday: 5 students
Wednesday: 2 students
Thursday: 3 students

A) The median will stay the same.

B) The median will be smaller.

C) The median will be larger.

---

1. If 0 students walk to the school on Friday, how will the new median compare?

2. If 1 student walks to the school on Friday, how will the new median compare?

3. If 2 students walk to the school on Friday, how will the new median compare?

4. If 3 students walk to the school on Friday, how will the new median compare?

5. If 4 students walk to the school on Friday, how will the new median compare?

Notes Template: QUEST Cycle

*Q*uestioning student understanding of a particular concept

Considerations: What is the concept you wish to target? Is the concept at grade level, or is it a prerequisite?

_____

_____

_____

_____

_____

_____

*U*ncovering understandings and misunderstandings using a probe

Considerations: How will you collect information from students (i.e., paper and pencil, interview, student response system)? What form will you use (i.e., one-page probe, card sort). Are there adaptations you plan to make? Review the summary of typical student responses. What do you predict to be common understandings and/or misunderstandings for your students?

_____

_____

_____

_____

_____

_____

*E*xamining student work

Sort by selected responses then resort by trends in thinking.

Considerations: What common understanding and misunderstandings were elicited by the probe?

_____

_____

_____

_____

_____

_____

*S*eeking links to cognitive research to drive next steps in instruction

Considerations: How do these elicited understanding and misunderstandings compare to those listed in the Teacher Notes? Review the bulleted items in the Focus Through Instruction and Questions to Consider to begin planning next steps. What additional sources did you review?

_____

_____

_____

_____

_____

_____

*T*eaching implications based on findings and determining impact on learning by asking an additional question

Considerations: What actions did you take? How did you assess the impact of those actions? What are your next steps?

_____

_____

_____

_____

_____

_____

_____

References

American Association for the Advancement of Science. (1989). *Science for all Americans: A project 2061 report on literacy goals in science, mathematics, and technology.* Washington, DC: Author.

American Association for the Advancement of Science. (1993). *Benchmarks for science literacy.* New York: Oxford University Press.

Anthony, G. J., & Walshaw, M. A. (2004). Zero: A "none" number? *Teaching Children Mathematics, 11*(1), 38–42.

Askew, M., & Wiliam, D. (1995). *Recent research in mathematics education 5–16.* London: HMSO Publications.

Assessment Reform Group. (2002). *Assessment for learning: 10 principals* [Pamphlet]. London: Author.

Bay Area Mathematics Task Force. (1999). *A mathematics sourcebook for elementary and middle school teachers.* Novato, CA: Arena Press.

Berkas, N., & Pattison, C. (2008, April). Differentiated instruction and universal design for learning. *NCTM News Bulletin.* National Council of Teachers of Mathematics. Retrieved May 25, 2010, from http://www.nctm.org/news/release_list.aspx?id=14867.

Bredekamp, S. (2004). Play and school readiness. In E. F. Zigler, D. G. Singer, & S. J. Bishop-Josef (Eds.), *Children's play the roots of reading* (pp.159–174). Washington, DC: Zero to Three Press.

Bright, G., & Joyner, J. (2004). *Dynamic classroom assessment: Linking mathematical understanding to instruction.* Vernon Hills, IL: ETA/Cuisenaire.

Burns, M. (2000). *About teaching mathematics: A K–8 resource.* Sausalito, CA: Math Solutions.

Burns, M. (2005). Looking at how students reason. *Educational Leadership: Assessment to Promote Learning, 63*(3), 26–31.

Clarke, D. M., Roche, A., & Mitchell, A. (2008). 10 Practical tips for making fractions come alive and make sense. *Mathematics Teaching in the Middle School, 13*(7), 372–379.

Clements, D., & Sarama, J. (2004). *Engaging young children in mathematics: Standards for early childhood mathematics education.* Mahwah, NJ: Lawrence Erlbaum.

DuFour, R., DuFour, R, Eaker, R., & Many, T. (2006). *Learning by doing: A handbook for professional learning communities at work.* Bloomington, IN: Solution Tree.

Fuson, K. C., Wearne, D., Hiebert, J., Human, P., Murray, H., Olivier, A., et al. (1997). Children's conceptual structures for multidigit numbers and methods of multidigit addition and subtraction. *Journal for Research in Mathematics Education, 28*(2), 130–162.

Geary, D. C. (1994). *Children's mathematical development: Research and practical Applications.* Washington, DC: American Psychological Association.

Goldman, J. [Ed]. (2000). Publishing definition on p. 257. Cleveland, OH: Wiley.

Griffin, P., & Madgwick, S. (2005). *Multiplication makes bigger and other mathematical myths.* Sowton, UK: DCS Publications.

Keeley, P., & Rose, C. M. (2006). *Mathematics curriculum topic study: Bridging the gap between standards and practice.* Thousand Oaks, CA: Corwin.

Leahy, S., Lyon, C., Thompson, M., & Wiliam, D. (2005). Classroom assessment: Minute by minute, day by day. *Educational Leadership: Assessment to Promote Learning, 63*(3), 19–24.

Lindquist, M., & Joyner, J. M. (2004). Mathematics guidelines for preschool. In D. H. Clements, A. M. DiBiase, & J. Sarama (Eds.), *Engaging young children in mathematics: Standards for early childhood mathematics education.* Mahwah, NJ: Lawrence Erlbaum.

Loucks-Horsley, S., Love, N., Stiles, K., Mundry, S., & Hewson, P. (2003). *Designing professional development for teachers of science and mathematics.* Thousand Oaks, CA: Corwin.

Mack, N. (1990, January). Learning fractions with understanding: Building on informal knowledge. *Journal for Research in Mathematics Education, 21,* 16–32.

McTighe, J., & O'Conner, K. (2005). Seven practices for effective learning. *Educational Leadership: Assessment to Promote Learning, 63*(3), 10–17.

McREL. (2002). *EDThoughts: What we know about mathematics teaching and learning.* Bloomington, IN: Solution Tree.

Mestre, J. (1989). Hispanic and Anglo students' misconceptions in mathematics. *ERIC Digest.* Appalachia Educational Laboratory. (ERIC Document Reproduction Service No. ED313192)

Minton, L. (2007). *What if your ABCs were your 123s? Building connections between literacy and numeracy.* Thousand Oaks, CA: Corwin.

National Council of Teachers of Mathematics. (1993a). *Research ideas for the classroom: Early childhood mathematics.* Reston, VA: Author.

National Council of Teachers of Mathematics. (1993b). *Research ideas for the classroom: Middle grades mathematics.* New York: Macmillan.

National Council of Teachers of Mathematics. (2000). *Principles and standards for school mathematics.* Reston, VA: Author.

National Council of Teachers of Mathematics. (2000–2010). *Illuminations.* Retrieved May 15, 2010, from http://illuminations.nctm.org.

National Council of Teachers of Mathematics. (2002a). *Lessons learned from research.* Reston, VA: Author.

National Council of Teachers of Mathematics. (2002b). *Putting research into practice in the elementary grades: Readings from journals of the NCTM.* Reston, VA: Author.

National Council of Teachers of Mathematics. (2002c). *Reflecting on NCTM's principles and standards in elementary and middle school mathematics.* Reston, VA: Author.

National Council of Teachers of Mathematics. (2003). *Research companion to principles and standards for school mathematics.* Reston, VA: Author.

National Council of Teachers of Mathematics. (2006). *Curriculum focal points for prekindergarten through grade 8 mathematics: A quest for coherence.* Reston, VA: Author.

National Research Council. (2001). *Adding it up: Helping children learn mathematics.* Washington, DC: National Academy Press.

National Research Council. (2002). *Helping children learn mathematics.* Washington, DC: National Academy Press.

National Research Council. (2005). *How students learn: Mathematics in the classroom.* Washington, DC: National Academy Press.

Naylor, S., & Keogh, B. (2000). *Concept cartoons in science education.* Sandbach, UK: Millgate House Education.

Paulos, J. A. (1991). *Beyond numeracy.* New York: Vintage.

Popsicle® is a registered trademark of Unilever Supply Chain, Inc.

Resnick, L., & Omanson, S. (1987). Learning to understand arithmetic. In R. Glaser (Ed.), Advances in instructional psychology (Vol. 3, pp. 41–95). Hillsdale, NJ: LEA.

Reynolds, A. (1993). School Effectiveness: Research, Policy and Practice. *School Effectiveness and School Improvement: An International Journal of Research, Policy and Practice, 4*(4), 311-317. doi:10.1080/0924345930040405

Rose, C., & Arline, C. (2009). *Uncovering student thinking in mathematics, grades 6–12: 30 formative assessment probes for the secondary classroom.* Thousand Oaks, CA: Corwin.

Rose, C., Minton, L., & Arline, C. (2007). *Uncovering student thinking in mathematics: 25 formative assessment probes.* Thousand Oaks, CA: Corwin.

Senge, P., Kleiner, A., Roberts, C., Smith, B., & Ross, R. (1994). *The fifth discipline fieldbook.* New York: Doubleday.

Shulman, L. S. (1987). Knowledge and teaching: Foundations of the new reform. *Harvard Educational Review, 57*(1), 1–22.

Sowder, J. L., & Nickerson, S. (2010). *Reconceptualizing mathematics for elementary school teachers.* Boston: W. H. Freeman.

Stepans, J. I., Schmidt, D. L., Welsh, K. M., Reins, K. J., Saigo, B. W. (2005). *Teaching for K–12 mathematical understanding using the conceptual change model.* St. Cloud, MN: Saiwood.

Taylor-Cox, J. (2008). *Differentiating in number and operations.* Portsmouth, NH: Heinemann.

Tomlinson, C. (1995). *How to differentiate instruction in mixed-ability classrooms.* Alexandria, VA: Association for Supervision and Curriculum Development.

Tomlinson, C. A. (2001). *How to differentiate instruction in mixed-ability classrooms.* (2nd ed.). Alexandria, VA: Association for Supervision and Curriculum Development.

University of Kansas. (2005). *Error pattern analysis.* Retrieved May 29, 2010, from http://www.specialconnections.ku.edu/~specconn/page/instruction/math/pdf/patternanalysis.pdf.

Van de Walle, J. A. (2007). *Elementary and middle school mathematics* (6th ed.). Boston: Pearson.

Van de Walle, J. A. (2010). *Elementary and middle school mathematics* (7th ed.). Boston: Pearson.

Walsh, J. & Sattes, B. (2005). *Quality questioning research-based practice to engage every learner.* Thousand Oaks: Corwin and AEL.

Yetkin, E. (2003). Student difficulties in learning elementary mathematics. *Eric Digest.* ERIC Clearinghouse for Science Mathematics and Environmental Education. (Document Reproduction Service No. ED482727)

Index

Action research, 17
  See also QUEST cycle
Age-related goals, 30–31
Algebraic thinking:
  equality concept and, 37, 80
  part-part-whole number concepts and, 33
American Association for the Advancement of Science
    (AAAS), 7, 8, 20, 43, 50, 60, 73, 74, 105, 107,
    111, 117, 130, 138, 146, 148, 152, 153, 157,
    158, 162, 163, 168, 189, 190, 196, 197
Anthony, G. J., 112, 113
Applets, 22, 51, 61, 66, 81, 88, 94, 102, 119, 125,
    169, 182
Arline, C., 3, 17, 22, 45, 52, 57, 62, 67, 74, 82, 89,
    95, 103, 107, 113, 120, 126, 132, 140, 148,
    153, 163, 169, 177, 183, 190, 197
Assessment initiatives, vii, ix, 40
  comprehensive diagnostic assessment systems, ix
  formative assessments and, viii, ix
  local/state assessments, 37
  point of entry assessment, 30–31
  pre-/postassessment format, 38
  prior knowledge, assessment of, 31
  probes and, vii, 1, 2 (figure)
  student preconceptions and, 40
  summative assessments and, ix, 7
  See also Diagnostic assessment; Mathematics
    Assessment Probes
Assessment Reform Group, 29

Bay Area Mathematics Task Force, 86, 93, 101, 103,
    113, 148, 153, 183
Berkas, N., 30
Booker, G., 112
Bredekamp, S., 31
Bright, G., 33
Burns, M., 31, 45, 52, 67

Calculation Nation (NCTM), 88, 102
Capacity building focus, 38–39
Clarke, D. M., 60
Clements, D., 51, 52
Cognitive research, 18, 19, 20, 20 (figure), 24
Common error patterns, 5–6, 5 (figure), 10, 19
Communication skill development, 34, 36–37

Comprehensive diagnostic assessment systems, ix
Computation/estimation, 115 (figure)
  How Many Dots? probe, 116–121
  Is It an Estimate? probe, 155–159
  part-part-whole number concepts and, 33
  Play Ball probe, 122–127
  What Is Your Estimate? probe, 160–164
  What's Your Addition Strategy? probe, 128–135
  What's Your Division Strategy? probe, 150–154
  What's Your Multiplication Strategy? probe,
    144–149
  What's Your Subtraction Strategy? probe, 136–143
Conceptual understanding, 2–4, 4 (figure), 10, 19
Content knowledge, viii, 29
Crayon Count probe, 24 (figure), 26 (figure),
    77 (figure)
  Examining student work and, 86
  focus through instruction and, 88
  Questioning for student understanding and, 86
  questions to consider and, 88
  references for, 89
  Seeking links to cognitive research and, 86–87
  student interview task and, 85
  student responses to, 89
  teacher sound bite and, 88
  Teaching implications and, 87
  Uncovering understanding and, 86
  variation on, 90
  See also Parts/wholes/equality
Curriculum Topic Study (CTS) process,
    21, 22, 39
Curriculum Topic Study Project, x

Data. *See* Measurement/geometry/data
Diagnostic assessment, vii, ix, 1, 29
  probes for, 1, 2 (figure)
  See also Assessment initiatives; Mathematics
    Assessment Probes
Differentiated instruction, 30, 31
Distractors, 10, 19
Duckworth, E., viii
DuFour, R., 39

Eaker, R., 39
Early Mathematics Thinking Project, x

Educational Development Center, 22
Effect on the Median probe, 198
Elaboration tier, 15–17
    See also Elicitation tier; Mathematics Assessment
        Probes
Elicitation tier, 10
   distractors and, 10
   examples/nonexamples list category,
        14, 14 (figure), 35
   justified list category, 14, 15 (figure)
   multiple selections response category, 12, 12 (figure)
   open response category, 12, 13 (figure)
   opposing views/answers category, 13, 14 (figure)
   selected response category, 11, 11 (figures)
   strategy elicitation category, 15, 16 (figure)
   See also Elaboration tier; Mathematics Assessment
        Probes
Equal to Four? probe, 14 (figure), 24 (figure),
        26–27 (figures), 37, 77 (figure)
   Examining student work and, 79
   focus through instruction and, 80–81
   Questioning for student understanding and, 79
   questions to consider and, 81
   references for, 82
   Seeking links to cognitive research and, 79–80
   student interview task and, 78
   student responses to, 82
   teacher sound bite and, 81
   Teaching implications and, 80
   Uncovering understanding and, 79
   variations on, 83–84
   See also Parts/wholes/equality
Equal to Fourteen? probes, 83–84
Equality. *See* Parts/wholes/equality
Error patterns, 5–6, 5 (figure), 10
Examples/nonexamples list category, 14, 14 (figure), 35
Explanation prompts. *See* Elaboration tier

Feghali, I., 111
Formative assessments, viii, ix
   See also Assessment initiatives
Fuson, K. C., 44

Geary, D. C., 31
Generalizations:
   composition/decomposition of numbers and, 34
   overgeneralization error, 6–7, 6 (figure)
Geometry. *See* Measurement/geometry/data
Goals, 38
Granola Bar probe, 24 (figure), 27 (figure), 77 (figure)
   Examining student work and, 100
   focus through instruction and, 101–102
   Questioning for student understanding and, 100
   questions to consider and, 102
   resources for, 103
   Seeking links to cognitive research and, 100–101
   student interview task and, 99
   student responses to, 103
   teacher sound bite and, 102
   Teaching implications and, 101

   Uncovering understanding and, 100
   See also Parts/wholes/equality
Graph Choices probe, 25 (figure), 27 (figure),
        165 (figure)
   Examining student work and, 188
   focus through instruction and, 189
   Questioning for student understanding and, 188
   questions to consider and, 189–190
   references for, 190
   Seeking links to cognitive research and, 188–189
   student interview task and, 187
   student responses to, 191
   teacher sound bite and, 190
   Teaching implications and, 189
   Uncovering understanding and, 188
   variations on, 192–193
   See also Measurement/geometry/data
Gumballs-in-a-Jar exercise, 7–10, 8–9 (figures)

Hewson, P., 17, 39
Hiebert, J., 44
Hinge points, 36
How Many Counters? probe, 26 (figure), 48, 121
How Many Dots? probe, 25–26 (figures), 33, 115 (figure)
   Examining student work and, 117
   focus through instruction and, 119
   Questioning for student understanding and, 117
   questions to consider and, 119
   resources for, 120
   Seeking links to cognitive research and, 117–119
   student interview task and, 116
   student responses to, 120
   teacher sound bite and, 120
   Teaching implications and, 119
   Uncovering understanding and, 117
   variation on, 121
   See also Computation/estimation
How Many Stars? probe, 13 (figure), 24 (figure),
        26 (figure), 41 (figure)
   Examining student work and, 43
   focus through instruction and, 44–45
   Questioning for student understanding and, 43
   questions to consider and, 45
   references for, 45
   Seeking links to cognitive research and, 43–44
   student interview task and, 42
   student responses to, 46–47
   teacher sound bite and, 45
   Teaching implications and, 44
   Uncovering understanding and, 43
   variation on, 48
   See also Place value/number charts/number lines
How Much Is Shaded? probe, 98
Human, P., 44
Hundred Chart Chunks probe, 24 (figure), 26 (figure),
        38, 41 (figure)
   Examining student work and, 65
   focus through instruction and, 66
   Questioning for student understanding and, 65
   questions to consider and, 67

resources for, 67
Seeking links to cognitive research and, 65–66
student interview task and, 64
student responses to, 68–69
teacher sound bite and, 67
Teaching implications and, 66
Uncovering understanding and, 65
variation on, 70
See also Place value/number charts/number lines

Illuminations (NCTM), 51, 81, 119, 125, 169
Images from Practice:
How Many Dots? probe, 33
Hundred Chart Chunk probe, 38
Is It Equal to 4? probe, 37
Play Ball probe, 34
Using Probes Across the District, 39
What Is the Value of the Place? probe, 35
See also Instructional implications; Place
value/number charts/number lines
Instructional implications, vii, ix, x, 2, 29
capacity building focus across grades/spans, 38–39
communication skill development, 34, 36–37
Curriculum Topic Study process and, 21
differentiated instruction and, 30, 31
hinge points in lessons and, 36
individual think time, opportunities for, 35–36
instructional activities/materials effectiveness,
assessment of, 37–38
instructional level, ascertainment of, 31
math conversations, promotion of, 33–34
Mathematics Assessment Probes, purposes of,
29–30
point of entry, assessment of, 30–31
process skill development, 36–37
QUEST cycle and, 21, 21–22 (figure)
student interviews, opportunities for, 32–33
student preconceptions, explicit attention to, 32, 40
student thinking trends, analysis of, 31–32, 33
vocabulary, development of, 34–35
See also Probes
Interactive technology applets, 22, 51, 61, 66, 81, 88,
94, 102, 119, 125, 169, 182
Internet resources. *See* Interactive technology applets
Interviews. *See* Student interviews
Is $\frac{1}{4}$ of the Whole Shaded? probe, 24 (figure),
26–27 (figures), 77 (figure)
Examining student work and, 92–93
focus through instruction and, 94
Questioning for student understanding and, 92
questions to consider and, 94
references for, 95
Seeking links to cognitive research and, 93–94
student interview task and, 91
student responses to, 95–96
teacher sound bites and, 95
Teaching implications and, 94
Uncovering understanding and, 92
variations on, 97–98
See also Parts/wholes/equality

Is It an Estimate? probe, 25 (figure), 27 (figure)
Examining student work and, 156
focus through instruction and, 157–158
Questioning for student understanding and, 156
questions to consider and, 158
resources for, 158
Seeking links to cognitive research and, 156–157
student interview task and, 155
student responses to, 159
teacher sound bite and, 158
Teaching implications and, 157
Uncovering understanding and, 156
See also Computation/estimation
Is It a Circle? probe, 26 (figure), 173
Is It Equivalent? probe, 15 (figure), 24 (figure),
27 (figure), 77 (figure)
Examining student work and, 105
focus through instruction and, 206
Questioning for student understanding and, 105
questions to consider and, 106–107
references for, 107
Seeking links to cognitive research and, 105–106
student interview task and, 104
student responses to, 108
teacher sound bite and, 107
Teaching implications for, 106
Uncovering understanding and, 105
variation on, 109
See also Parts/wholes/equality
Is It a Polygon? probe, 26–27 (figures), 172
Is It Simplified? probe, 24 (figure), 27 (figure),
77 (figure)
Examining student work and, 111
focus through instruction and, 112
Questioning for student understanding and, 111
questions to consider and, 112
references for, 113
Seeking links to cognitive research and, 111–112
student interview task and, 110
student responses to, 113
teacher sound bite and, 112
Teaching implications and, 112
Uncovering understanding and, 111
See also Parts/wholes/equality

Joyner, J. M., 31, 33
Justification methods, 37
Justified list category, 14, 15 (figure)

Keeley, P., 2, 7, 8, 9, 21, 22, 39, 45, 52, 57, 67, 74, 82,
89, 95, 103, 107, 113, 120, 126, 132, 140, 148,
153, 158, 163, 169, 177, 183, 190, 197
Keogh, B., 13
Kleiner, A., 38

Large-scale diagnostic assessment systems, ix
Leahy, S., 36
Learning. *See* Formative assessments; Instructional
implications; Mathematics Assessment Probes;
Student learning

Lindquist, M., 31
Loucks-Horsley, S., 17, 39
Love, N., 17, 39
Lyon, C., 36

Mack, N., 86
McREL, 132, 140
Maine Governor's Academy for Mathematics and
    Science Education Leadership, x
Maine Mathematics and Science Alliance, x
Many, T., 39
Mathematical literacy, vii, 37, 38
Mathematics Assessment Probes, vii, ix, 1
    common error patterns and, 5–6, 5 (figure), 10
    conceptual understanding and, 2–4, 4 (figure), 10
    constructive feedback and, 7
    development process for, 7–10
    diagnostic assessment probes and, 1, 2 (figure)
    examples/nonexamples list category and, 14, 14
        (figure)
    grade-level view of, 25, 26–27 (figure)
    instructional cycle for, 2
    interactive technology applets and, 22
    justified list category and, 14, 15 (figure)
    misunderstandings, elicitation of, ix, x, 1–2, 4–7,
        5–6 (figures)
    multiple selections response category and,
        12, 12 (figure)
    national standards and, 7, 8 (figure), 24
    open response category and, 12, 13 (figure)
    opposing views/answers category and, 13, 14 (figure)
    overgeneralizations and, 6–7, 6 (figure)
    piloting developed probes and, 7
    probability example, 7, 8–9 (figures), 10
    purposes of, 29–30
    research findings and, 7, 8 (figure)
    research-based tools of, x–xi
    selected response category and, 11, 11 (figures)
    strategy elicitation category and, 15, 16 (figure)
    structure of, 10–17
    summary table of probes, 24, 24–25 (figure)
    targeted instructional choices and, ix, x, 1, 2
    Tier 1/elicitation tier, categories of, 10–16,
        11–16 (figures)
    Tier 2/elaboration tier and, 15–17
    uncommon misunderstandings and, 7
    understandings, elicitation of, ix, x, 1, 2–4, 4 (figure)
    variation/adaptation of the probes and,
        18, 19 (figure), 25, 26–27 (figure)
    See also Instructional implications; QUEST cycle
Mathematics Curriculum Topic Study, 39
Measurement errors, 5–6, 5 (figure)
Measurement/geometry/data, 165 (figure)
    Graph Choices probe, 187–193
    The Median probe, 194–198
    Quadrilaterals probe, 166–173
    What's the Area? probe, 174–179
    What's the Measure? probe, 180–186
The Median probe, 25 (figure), 27 (figure), 165 (figure)
    Examining student work and, 195
    focus through instruction and, 196

Questioning for student understanding and, 195
questions to consider and, 196
references for, 197
Seeking links to cognitive research and, 195–196
student interview task and, 194
student responses to, 198
teacher sound bite and, 196
Teaching implications and, 196
Uncovering understanding and, 195
variation on, 198
See also Measurement/geometry/data
Mestre, J., 4
Metacognition strategies, 34
Milestones, 38
Minton, L., 3, 4, 9, 11, 15, 17, 23, 34, 42, 45, 48, 49,
    52, 53, 54, 57, 58, 62, 63, 64, 67, 70, 71, 74, 76,
    78, 82, 85, 89, 90, 91, 95, 97, 98, 99, 103, 104,
    109, 110, 116, 118, 120, 121, 122, 126, 127,
    128, 134, 135, 136, 144, 149, 150, 155, 160,
    163, 166, 169, 174, 177, 179, 180, 183, 187,
    190, 194, 198, 200
Misconceptions. *See* Mathematics Assessment Probes;
    Misunderstandings
Misunderstandings, ix, x, 1–2, 4–5
    common error patterns, 5–6, 5 (figure)
    overgeneralizations, 6–7, 6 (figure)
    student preconceptions, explicit attention to, 32
    See also Mathematics Assessment Probes
Mitchell, A., 60
Multiple selections response category, 12, 12 (figure)
Mundry, S., 17, 39
Murray, H., 44

Name the Graph probe, 192
Name That Shape probe, 171
National Council of Teachers of Mathematics (NCTM),
    2, 6, 7, 8, 10, 20, 22, 24, 33, 36, 37, 43–44, 45,
    50, 51, 52, 55–56, 57, 60, 62, 65–66, 67, 73, 74,
    80, 82, 87, 89, 93, 94, 95, 100–101, 103, 106,
    107, 113, 118, 119, 120, 124, 126, 130, 131,
    132, 138, 139, 140, 146, 148, 152, 153, 156,
    157, 158, 162, 163, 168, 169, 176, 177, 182,
    183, 189, 190, 195, 196, 197
National Library of Virtual Manipulatives,
    22, 61, 66
National Research Council (NRC), 20, 40, 45, 65, 67,
    82, 93, 106, 107, 130, 131, 132, 138, 139, 140,
    148, 152, 153, 154, 169, 183
National Science Foundation (NSF), viii, x
Naylor, S., 13
Nickerson, S., 56, 57, 60, 62
No Child Left Behind Act of 2001, ix
Northern New England Co-Mentoring Network, x
Number charts. *See* Place value/number charts/
    number lines
Number lines. *See* Place value/number charts/
    number lines

Olivier, A., 44
Omanson, S., 124
Open response category, 12, 13 (figure)

Opposing views/answers category, 13, 14 (figure)
Overgeneralizations, 6–7, 6 (figure), 19

Parts/wholes/equality, 77 (figure)
    Crayon Count probe, 85–90
    Equal to Four? probe, 78–84
    Granola Bar probe, 99–103
    Is $\frac{1}{4}$ of the Whole Shaded? probe, 91–98
    Is It Equivalent? probe, 104–109
    Is It Simplified? probe, 110–113
Pattison, C., 30
Paulos, J. A., 20, 132
Place value/number charts/number lines, 41 (figure)
    How Many Stars? probe, 42–48
    Hundred Chart Chunks probe, 64–70
    What Is the Value of the Digit? probe, 71–76
    What Is the Value of the Place? probe, 35, 54–57
    What Number Is That? probe, 58–63, 73
    What's the Number? probe, 49–53
Play Ball probe, 25–26 (figures), 34, 115 (figure)
    Examining student work and, 123
    focus through instruction and, 124–125
    questions to consider and, 125
    Questioning for student understanding and, 123
    resources for, 126
    Seeking links to cognitive research and, 123–124
    student interview task and, 122
    student responses to, 126
    teacher sound bite and, 125
    Teaching implications for, 124
    Uncovering understanding and, 123
    variation on, 127
    See also Computation/estimation
Point of entry, 30–31
Prerequisite skills, ix
Prior knowledge, ix, 1, 31, 40
Probability concept, 7–10, 8–9 (figures)
Probes, vii, 1, 2 (figure)
    Crayon Count, 85–90
    Equal to Four?, 37, 78–84
    Granola Bar, 99–103
    Graph Choices, 187–193
    How Many Dots?, 33, 116–121
    How Many Stars?, 42–48
    Hundred Chart Chunks, 38, 64–70
    Is $\frac{1}{4}$ of the Whole Shaded?, 91–98
    Is It an Estimate?, 155–159
    Is It Equivalent?, 104–109
    Is It Simplified?, 110–113
    The Median, 194–198
    Play Ball, 34, 122–127
    Quadrilaterals, 166–173
    Using Probes Across the District, 39
    What Is the Value of the Digit?, 71–76
    What Is the Value of the Place?, 35, 54–57
    What Is Your Estimate?, 160–164
    What Number Is That?, 58–63
    What's the Area?, 174–179
    What's the Measure?, 180–186
    What's the Number?, 49–53
    What's Your Addition Strategy?, 128–135

    What's Your Division Strategy?, 150–154
    What's Your Multiplication Strategy?, 144–149
    What's Your Subtraction Strategy?, 136–143
    See also Instructional implications; Mathematics
        Assessment Probes
Procedural knowledge, 2–4, 4 (figure), 19
Process skill development, 36–37
Professional development, viii
    action research and, 17
    Using Probes Across the District probe, 39

Quadrilaterals probe, 14 (figure), 25 (figure),
    27 (figure), 165 (figure)
    Examining student work and, 167
    focus through instruction and, 168–169
    Questioning for student understanding and, 167
    questions to consider and, 169
    references for, 169
    Seeking links to cognitive research and, 167–168
    student interview task and, 166
    student responses to, 170
    teacher sound bite and, 169
    Teaching implications and, 168
    Uncovering understanding and, 167
    variations on, 171–173
    See also Measurement/geometry/data
QUEST cycle, 17, 17 (figure), 25
    Examining student work and, 19, 19 (figure)
    note-taking template for, 22, 23 (figure), 199–200
    Questioning for student understanding and,
        18, 18 (figure)
    Seeking links to cognitive research and,
        20, 20 (figure)
    Teaching implications and, 21, 21–22 (figure)
    Uncovering understanding and, 18, 19 (figure)
    See also Mathematics Assessment Probes; Probes
Questioning strategy, ix, 22
    hinge point questions and, 36
    individual think time and, 35–36
    traditional classroom questioning, 36
    wait time and, 35
    See also Instructional implications; QUEST cycle;
        Student interviews

Reins, K. J., 32, 177, 190, 197
Resnick, L., 124
Resources:
    interactive technology applets, 22, 51, 61, 66, 81,
        88, 94, 102, 119, 125, 169, 182
    research/teaching implications and, 22
Reynolds, A., 29
Roberts, C., 38
Roche, A., 60
Rose, C., 3, 7, 8, 9, 17, 21, 22, 39, 45, 52, 57, 62,
    67, 74, 82, 89, 95, 103, 107, 113, 120, 126,
    132, 140, 148, 153, 158, 163, 169, 177, 183,
    190, 197
Ross, R., 38

Saigo, B. W., 32, 177, 190, 197
Sarama, J., 51, 52

Sattes, B., 35
Schmidt, D. L., 32, 177, 190, 197
Schulman, L. S., 29
Selected response category, 11, 11 (figures)
Senge, P., 38
Smith, B., 38
Sowder, J. L., 56, 57, 60, 62
Standards:
    local/state assessments and, 37
    mathematical literacy and, 37
    Mathematics Assessment Probes and, 7, 8 (figure), 24
    QUEST cycle and, 18
State Mathematics and Science Partnership Project, x
STEM education, viii
Stepans, J. I., 32, 177, 190, 197
Stiles, K., 17, 39
Strategy elicitation category, 15, 16 (figure)
Student interviews, 32–33
Student learning:
    age-related goals and, 30
    barriers to, viii
    differentiated instruction and, 30, 31
    incorrect thinking, correction of, 32
    individual think time and, 35–36
    informal knowledge of mathematics and, 31
    learning timeline and, 30–31
    metacognition strategies, practice of, 34
    point of entry, assessment of, 30–31
    prior knowledge and, 1, 31, 40
    whole child development and, 31
    See also Formative assessments; Instructional
        implications; Mathematics Assessment Probes;
        QUEST cycle
Student self-assessment, ix
Summative assessments, ix, 7
    See also Assessment initiatives
Symbolic notation, 34–35

Target concepts/procedures, ix
Taylor-Cox, J., 45
Teachers:
    pedagogical content knowledge and, viii, 39
    See also Professional development
Teaching process. *See* Instructional implications;
    Mathematics Assessment Probes; QUEST cycle
Technology. *See* Interactive technology applets
Thompson, M., 36
Tiers. *See* Elaboration tier; Elicitation tier; Mathematics
    Assessment Probes
Tobey, C. R., viii, 4, 9, 11, 15, 23, 42, 48, 49, 53, 54,
    58, 63, 64, 70, 71, 76, 78, 85, 90, 91, 97, 98, 99,
    104, 109, 110, 116, 121, 122, 127, 128, 134,
    135, 136, 144, 149, 150, 155, 160, 166, 174,
    179, 180, 187, 194, 198, 200
Tomlinson, C. A., 30

Understandings, ix, x, 1–2
    conceptual understanding, 2–4, 4 (figure)
    partial understandings, 32
    procedural knowledge and, 2–4, 4 (figure)
    See also Mathematics Assessment Probes;
        QUEST cycle

Van de Walle, J. A., 60, 62, 74, 87, 89, 95, 106, 107,
    112, 124, 126, 162, 163, 168, 176, 177, 181,
    183, 188, 190, 197
Van Hiele, J. 168

Wait time, 35
Walsh, J., 35
Walshaw, M. A., 112, 113
Wearne, D., 44
Welsh, K. M., 32, 177, 190, 197
What Does the Graph Say? probe, 193
What Is the Value of the Digit? probe, 26–27 (figures),
    41 (figure)
    Examining student work and, 72
    focus through instruction and, 73–74
    Questioning for student
        understanding and, 72
    questions to consider and, 74
    references for, 74
    Seeking links to cognitive research and, 73
    student interview task and, 71
    student responses to, 75
    teacher sound bite and, 74
    Teaching implications and, 73
    Uncovering understanding and, 72
    variation on, 76
    See also Place value/number charts/number lines
What Is the Value of the Place? probe, 24 (figure),
    26 (figure), 35, 41 (figure)
    Examining student work and, 55
    focus through instruction and, 56
    Questioning for student understanding and, 55
    questions to consider and, 56
    references for, 557
    Seeking links to cognitive research and, 55–56
    student interview task and, 54
    student responses to, 57
    teacher sound bite and, 57
    Teaching implications and, 56
    Uncovering understanding and, 55
    See also Place value/number charts/number lines
What Is Your Estimate? probe, 25 (figure), 27 (figure),
    115 (figure)
    Examining student work and, 161
    focus through instruction and, 162–163
    Questioning for student understanding and, 161
    questions to consider and, 163
    resources for, 163
    Seeking links to cognitive research and, 162
    student interview task and, 160
    student responses to, 164
    teacher sound bite and, 163
    Teaching implications and, 162
    Uncovering understanding and, 161
    See also Computation/estimation
What Number Is That? probe, 24 (figure), 26 (figure),
    41 (figure)
    Examining student work and, 59
    focus through instruction and, 61
    Questioning for student understanding and, 59
    questions to consider and, 61
    resources for, 62

Seeking links to cognitive research and, 60
student interview task and, 58
student responses to, 63
teacher sound bite and, 61
Teaching implications and, 60
Uncovering understanding and, 59
variation on, 63
See also Place value/number charts/number lines
What's the Area? probe, 12 (figure), 25 (figure), 27 (figure), 165 (figure)
Examining student work and, 175–176
focus through instruction and, 176–177
Questioning for student understanding and, 175
questions to consider and, 177
references for, 177
Seeking links to cognitive research and, 176
student interview task and, 174
student responses to, 178
teacher sound bite and, 177
Teaching implications and, 176
Uncovering understanding and, 175
variation on, 179
See also Measurement/geometry/data
What's the Length of the Line? probe, 27 (figure), 185
What's the Measure? probe, 25 (figure), 26–27 (figures), 165 (figure)
Examining student work and, 181
focus through instruction and, 182
Questioning for student understanding and, 181
questions to consider and, 183
references for, 183
Seeking links to cognitive research and, 181–182
student interview task and, 180
student responses to, 184
teacher sound bite and, 183
Teaching implications and, 182
Uncovering understanding and, 181
variations on, 185–186
See also Measurement/geometry/data
What's the Number? probe, 24 (figure), 26 (figure), 41 (figure)
Examining student work and, 50
focus through instruction and, 51
Questioning for student understanding and, 50
questions to consider and, 51
references for, 52
Seeking links to cognitive research and, 50–51
student interview task and, 49
student responses to, 52
teacher sound bite and, 52
Teaching implications and, 51
Uncovering understanding and, 50
variation on, 53
See also Place value/number charts/number lines
What's Your Addition Strategy? probe, 25–27 (figures), 115 (figure)
Examining student work and, 129–130
focus through instruction and, 131
Questioning for student understanding and, 129
questions to consider and, 132

resources for, 132
Seeking links to cognitive research and, 130–131
student interview task and, 128
student responses to, 133
teacher sound bite and, 132
Teaching implications and, 131
Uncovering understanding and, 129
variations on, 134–135
See also Computation/estimation
What's Your Division Strategy? probe, 25 (figure), 27 (figure), 115 (figure)
Examining student work and, 151
focus through instruction and, 152
Questioning for student understanding and, 151
questions to consider and, 153
references for, 153–154
Seeking links to cognitive research and, 151–152
student interview task and, 150
student responses to, 154
teacher sound bite and, 153
Teaching implications and, 152
Uncovering understanding and, 151
See also Computation/estimation
What's Your Multiplication Strategy? probe, 16 (figure), 25 (figure), 27 (figure), 115 (figure)
Examining student work and, 145–146
focus through instruction and, 147
Questioning for student understanding and, 145
questions to consider and, 147
resources for, 148
Seeking links to cognitive research and, 146
student interview task and, 144
student responses to, 148
teacher sound bite and, 147
Teaching implications and, 146
Uncovering understanding and, 145
variation on, 149
See also Computation/estimation
What's Your Subtraction Strategy? probe, 25–27 (figures), 115 (figure)
Examining student work and, 137–138
focus through instruction and, 139
Questioning for student understanding and, 137
questions to consider and, 139–140
resources for, 140
Seeking links to cognitive research and, 138–139
student interview task and, 136
student responses to, 141
teacher sound bite and, 140
Teaching implications and, 139
Uncovering understanding and, 137
variations on, 142–143
See also Computation/estimation
Wheeler, M., 111
Whole child development, 31
Wholes. *See* Parts/wholes/equality
Wiliam, D., 36

Zero point, 5–6, 5 (figure)

CORWIN
A SAGE Company

The Corwin logo—a raven striding across an open book—represents the union of courage and learning. Corwin is committed to improving education for all learners by publishing books and other professional development resources for those serving the field of PreK–12 education. By providing practical, hands-on materials, Corwin continues to carry out the promise of its motto: **"Helping Educators Do Their Work Better."**